DEMING'S PROFOUND CHANGES

When Will the Sleeping Giant Awaken?

Kenneth T. Delavigne

J. Daniel Robertson

Kanigel
This book
AC's 5 yr Eng
EWT's Sci Mgmt
[Reason re latent
dsg
error

brooks - Mythical
M-M

PTR Prentice Hall
Englewood Cliffs, NJ 07632

Library of Congress Cataloging-In-Publication Data

```
Delavigne, Kenneth T.
     Deming's profound changes : when will the
     sleeping giant awaken? / by Kenneth T.
     Delavigne & J. Daniel Robertson.
       p.    cm.
     Includes bibliographical references and index.
     ISBN 0-13-292690-3
     1. Management.  2. Deming, W. Edwards (William
     Edwards), 1900-1993
     —Contributions in management.  I. Robertson,
     J. Daniel. II. Gluckman, Perry, 1938-1992.
     III. Title.
HD38.D439D45  1994                         93-42352
658—dc20                                        CIP
```

Editorial/production supervision: BooksCraft, Inc., Indianapolis, IN
Jacket design: Design Solutions
Jacket photo: The IMAGEBANK, stock illustrations © Curt Doty
Interior design: Dit Mosco and Robert Coalson
Acquisitions editor: Michael Hays
Manufacturing manager: Alexis R. Heydt

© 1994 by Kenneth T. Delavigne and J. Daniel Robertson

Published by PTR Prentice Hall
Prentice-Hall, Inc.
A Paramount Communications Company
Englewood Cliffs, NJ 07632

The publisher offers discounts on this book when ordered in
bulk quantities. For more information contact: Corporate
Sales Department, PTR Prentice Hall, 113 Sylvan Avenue,
Englewood Cliffs, NJ 07632, Phone: 201-592-2863, FAX:
201-592-2249

Printed in the United States of America
10 9 8 7 6 5 4 3 2 1

ISBN 0-13-292690-3

ISBN 0-13-292690-3
90000
9 780132 926904

Prentice-Hall International (UK) Limited, *London*
Prentice-Hall of Australia Pty. Limited, *Sydney*
Prentice-Hall Canada Inc., *Toronto*
Prentice-Hall Hispanoamericana, S.A., *Mexico*
Prentice-Hall of India Private Limited, *New Delhi*
Prentice-Hall of Japan, Inc., *Tokyo*
Simon & Schuster Asia Pte. Ltd., *Singapore*
Editora Prentice-Hall do Brasil, Ltda., *Rio de Janeiro*

DEDICATIONS

To Beverly, my wife and bringer of books and parchments. (KTD)

To Vi, my partner and most trusted advisor. (JDR)

Contents

Figures

Tables

Foreword

by W. Edwards Deming

The boundaries of quality are fixed by the producer. The customer does not generate ideas about the product or quality that he needs. He learns from the producer what product or service might please him.

The quality of a product or service is the responsibility of top management. This responsibility cannot be delegated. A product or service must have a market. Without a market, production comes to a halt.

This book is for people that are living under the tyranny of the prevailing style of management. The huge, long-range losses caused by this style of management have led us into decline. Most people imagine that the present style of management has always existed, and is a fixture. Actually, it is a modern invention—a prison created by the way in which people interact. This interaction afflicts all aspects of our lives—government, industry, education, health care.

We have grown up in a climate of competition between people, teams, departments, divisions, pupils, schools, universities. We have been taught . . . that competition . . . will solve our problems. Actually, competition, we see now, is destructive. What we need is cooperation and transformation to a new style of management in which everyone works together as a system, with the aim for everybody to win.

The aim of Ken Delavigne and Dan Robertson in this book is to explain the responsibilities of top management for quality.

Washington
30 October 1993

ix

Preface

About a dozen books on W. Edwards Deming and his philosophy have been written—why another one? Because, as good as most of these books are, we felt that certain basic questions remained unasked (and unanswered), and certain issues unaddressed.

Over almost ten years we experienced the lack of appreciation for Deming's philosophy. During that period of major industrial decline, companies large and small seemed to grasp at every "productivity" nostrum that came down the road, only to slide even further. Dr. Deming made clear, in a message of compelling logic, what was wrong and what had to be done—yet his message was largely ignored by corporate America. Thus, as the 1980s wore on, increasingly we witnessed the "expensive funerals" Deming had predicted for organizations that did not undergo what he calls *transformation*. We sadly concluded that the state of management was even worse at the end of the 1980s than when the decade began. But why?

One hypothesis, suggested by our mentor and friend Perry Gluckman, was that people hearing Deming's philosophy were largely unaware that they themselves were acting under the influence of some philosophy of management—one quite different from his. Therefore, they thought that, not liking his stern pronouncements they could reject him without commitment to *any* other philosophy. Extensive research has made it clear that this is a false assumption:

- Current western management practice can in fact be traced directly back to the nineteenth century and Frederick Taylor's dubious theory of scientific management, and back beyond that into the eighteenth century.

- Over the century since Taylor, his philosophy has been adulterated and corrupted many times, resulting in the philosophy of management so widely—and unconsciously—practised today, for which we have coined the term "neo-Taylorism."

Nothing is more important about an individual than his or her world-view, his or her philosophy of life, because this is the basis for all actions. Management today know little of their own world-view, or even what to call it. We have now named it and will describe it in detail in this book.

These opposing world-views, Deming's philosophy and that of neo-Taylorism, are locked in battle for control of the management of the Western world—yet few realize this. The philosophical level is where the action is: ideas which have no immediate or locally-perceived effect may nevertheless beget consequences, diffused in both time and place, that change even the courses of nations. Our challenge is to make the unseen visible.

As it was absent from other books, we decided that an exposition of neo-Taylorism was needed here. To this, we will issue a radical challenge, using Deming's philosophy as our weapon. We have now presented these themes to many groups, and are encouraged that they may be powerful arguments in the battle for minds. Others may now become aware of this battle and decide where they stand.

Another chief aim of our book is to present Deming's philosophy under different aspects and classifications to enhance understanding, following Deming's own observation that people learn in different ways. So you will find that the structure of this book is not borrowed from either Deming's system of profound knowledge or his fourteen points, but is based upon a new interpretation of his philosophy—one that arose from asking once again the question, "Where did it come from?"

The last aim we will mention here is derived from Deming's great stress on *learning*. Asked why he maintained his teaching role even into his tenth decade of life, often spending as little as one day a month at home, Deming replied that he could not afford to stop learning from his students. Learning is an essential part of his philosophy, but it is often treated separately, under discussions of the theory of knowledge. But learning is perhaps the primary aim of the Plan-Do-Study-Act (PDSA) cycle, first formulated during the 1930s in the work and writings of Deming's greatest mentor, Walter Shewhart. This business application of the scientific method holds that an organization continuously gains the new knowledge needed for improvement through iterations of the PDSA cycle. Thus, subsequent improvements of product or process are but effects of that new knowledge. We have tried, therefore, to feature learning and knowledge to a greater degree than have previous discussions.

Learning is, in Deming's philosophy, a prelude to change—change for the better. Yet, as the longshoreman philosopher Eric Hoffer said, change is an ordeal, even when it is to a better state. We recognize this, and balance the discussions of learning with parallel analyses of anticipated change.

The specific aims that we have outlined may seem to favor a rather specialized group of people as the target audience, such as philosophers or those interested in the history of the quality movement. On the contrary, our focus is aimed at *widening* the audience for Deming's philosophy, and heightening its acceptance and impact. We are trying to clear a massive intellectual logjam, not merely talk to a few collectors of obscure or interesting bits of wood.

Everyone has a world-view, and acts in accordance with it. In fact, the most reliable way to discern another's world-view is to observe that person's actions. Listening only to what people claim they believe is usually misleading, as few have ever critically examined their own actions to see what philosophy they are acting out.

The largely unexamined philosophy of management that is leading the Western world to economic disaster is neo-Taylorism. *Everyone* needs to know more about neo-Taylorism: where it came from, what its beliefs are, and most importantly, how Deming's philosophy—if adopted—could be an antidote for it. This is the transformation Deming so ardently desires and works toward.

Deming's Profound Changes was born of thought, discussion, and research. The only interview we conducted was with Dr. Eugene Grant, who at age 97 is four years older than Deming, and had worked with him teaching applied statistics during World War II. Integration of management philosophy with other fields such as economics, physics, and philosophy itself was essential in our effort. One of the great delights of writing this book has been the constant process of discovery as we pursued our avenues of research; we encountered almost no dead ends, and in most cases the deeper we dug the richer the veins of gold waiting to be mined. Joy of research! A determined effort was needed to call a halt long enough to get our "final" manuscript in to Prentice Hall on the day appointed—and even after that the research notes kept piling up in hopes that we could add them in, which we did.

A philosophical transformation can follow no foreordained process, so we knew we didn't want to write a handbook or an implementation manual. Little is here on Deming the man, not even his picture, although we did feel it important to trace some of the more important events in his life. This is not a book about statistics; people like Shewhart, Deming, and Wheeler have already said everything we could say and far more—what we know, we learned from them. We don't merely regurgitate what Deming has already said quite eloquently himself; we assume that you have already, or will, read his own great books and attend his seminars for that. The study of organizational examples of "success" is without profit, so we included no distracting company profiles, interviews with CEOs, or before-and-after charts. Nor is this the "last Deming book"—but it *is* needed now.

We cordially invite you into the discussion that follows in this book!

ACKNOWLEDGMENTS

No book is written by the author or authors alone. Each is to some extent fostered by things that have gone before, thoughts that have been expressed, people who have taught and inspired. To rephrase a saying, no book is an island. Instead, it is the product of a complex chain of causality no one could hope to understand.

Of late we have become connoisseurs of the Acknowledgments sections of books, examining the traditional and novel ways authors have found to give some of the credit—and rarely any of the blame—to those who helped. The problem, of course, is order. Were we blissfully ignorant of Deming's system of profound knowledge, we would try to rank the people into some kind of pecking order according to degree of contribution, leaving each to ponder why his name was in *that* position in the list. But we know this is impossible. Not only would the list of names be longer than the book itself, but no system of ranking could give a "true" order. It would only be the result of a procedure. The procedure we have chosen here is simply an ordering by time within each grouping.

Contributors: Dave Buck, Siltec Silicon, Salem, Oregon; Bill Howell and Page Shirtum, Dow Chemical, Freeport, Texas; Masatoshi Hirai; Dr. W. Edwards Deming; Dr. George Watson, Syntex Corporation, Palo Alto; Dr. Eugene L. Grant, Stanford University (Ret.); Professor Israel M. Kirzner, New York University.

Reviewers: Beverly K. Delavigne; Vi Robertson; Marv Dettloff, Dow Chemical; Dr. W. Edwards Deming; Joanne Wojtusiak, Skylight Communications, Cornwall Bridge, Connecticut; Dr. A. V. Viswanathan, Boeing Aircraft, Seattle; Professor William J. Haga, U.S. Naval Postgraduate School, Monterey; Tim Fuller, Fuller Associates, Palo Alto; John Dowd, Santa Cruz; Joe Reid, Electroglas Corporation, Santa Clara; Professor Donald Gause, SUNY-Binghamton; Dr. Suzanne Godown, SUNY-Binghamton; Professor Edward T. Hall, Northwestern University; Dr. Henry Neave, President, British Deming Association.

Home Cheering Section: members of the Bay Area Deming Users Group and Deming Study Group; Dr. William R. Downing; members of the Golden Gate ASTD TQM professional practice group.

Suppliers: Apple Computer for making the Macintosh intuitive and even fun to use; and Microsoft for a Word processor that does almost everything we need. Various libraries—and librarians—too numerous to mention made countless invaluable, and sometimes obscure, works available to our hungry eyes.

IN MEMORIAM
Perry M. Gluckman, Ph.D.
1938–1992

A native of New York City, Perry Gluckman as a young boy moved to California with his parents. His formal education culminated in a doctorate in mathematical statistics from Stanford University, followed by a fellowship with statistician John Tukey. Later, his work at Lawrence Livermore Laboratories brought him into contact with one of W. Edwards Deming's wartime students, who encouraged Perry to apply Deming's methods to some of their problems.

It was during this effort that Perry came into regular contact with Deming himself and became one of his disciples. Setting up his own consulting firm, he served the manufacturing and process industries. Perry wrote numerous papers and articles, and in 1989 published the book *Everyday Heroes*, a set of six parables about real people who had decided to make a difference guided by various aspects of Deming's philosophy. He was a founding member of the Bay Area Deming Users Group.

Six months before his death in January 1992, Perry shared with a few of his friends the outline of what he envisioned as a new presentation advocating W. Edwards Deming's profound changes to the philosophy of management. Perry was uniquely—we choose that word with care—qualified for this role, and we were all eyes and ears as he talked us through the thirty-six slides he had put together. Although no one, least of all Perry, realized it at the time, his presentation, entitled *Deming's Profound Changes*, would become the matrix upon which this book was built.

It was not Perry's custom to do his thinking in isolation, and he welcomed the help and ideas of several allies, most of whom had met him in the course of his consulting practice. It soon was evident that *Deming's Profound Changes* would have to become something more than a slide presentation: a short explanation of main points would be added at the bottom of each transparency, making the package usable by someone other than Perry.

We followed this pattern for several months. Perry, bedridden from his recent operation for spinal cancer, was largely limited to the role of mentor and reviewer. *Profound Changes* slowly increased in size as we discussed the topics with Perry, wrote sections at home, and worked with him nights and weekends to enhance the document. We read many books from Perry's library, discussed subjects from his three-day course, and clipped newspaper and magazine articles.

It was not until Thanksgiving that we finally convinced Perry that *Profound Changes* was too important to remain a series of presenters' notes, and would have to become a book. Unaware that he had only a little more than eight remaining weeks to live, we stepped up our pace from then until almost Christmas. Christmas travel interrupted our work for over two weeks. When we got together to share what we had developed over the holidays, Perry was noticeably weaker and less able to concentrate.

Each of us unknowingly had a final session alone with Perry; we found out later that he had given each of us a list of topics that he felt needed to be added to complete the book. These, plus the themes and developments we originated, would, over the course of the next year and a half, increase more than fifteenfold the size of the last manuscript Perry saw. Some of the last words we remember are his declaration that he wouldn't be alive to see the book finished. A few days later, Perry Gluckman's battle with cancer was over.

It is a tribute to the job Perry did as teacher and mentor that in only a few cases have we felt an urgent need for further portions of his wisdom. His simple style of teaching without putting himself on a pedestal was an effective complement to a brilliant university-trained mind.

Which brings us to the variation in writing styles that you will encounter in this book, as in any book truly written by more than one author. Perry's style is easy to spot because it's conversational—he wrote as though he were speaking, sometimes using the first person. Although we helped him to enhance his text and ideas during those final several months, we have often included Perry's own words in this book because "no one says it better."

We acknowledge here Perry's inspiration, wisdom, and character. May Deming advocates everywhere profit from his influence upon our work.

IN MEMORIAM
W. Edwards Deming
1900–1993

At about three o'clock in the morning of 20 December, W. Edwards Deming passed away in his sleep. Death overtook him during one of those rare interludes between teaching and consulting trips when he could spend a little time at his home in Washington.

Although it would be difficult to identify any one person as "the father of quality," Deming was recognized round the world as the one who gave the modern quality movement the principles by which to think and act. His close disciples would go further than this, saying that his is a philosophy of life. Identifying him only with the quality movement would be for Deming a demotion.

Some reporters described Deming as a cult leader, a charismatic surrounded by unquestioning "true believers." Those who knew him knew otherwise: Deming, a man of simplicity and traditional stature, *lived* for questioning and learning from others.

Deming exhibited few characteristics of the cult leader: reticent almost to the point of shyness in public, he was unfailingly courteous to everyone, including people less qualified, and decades younger, than he. He demanded little of his followers, and returned more than they gave. He remembered the sources of his ideas—going back more than half a century—and gave them frequent credit in public. And unquestioning acceptance was the last thing he wanted: most of his close associates were professionals trained to doubt mere assertions of fact, and many of these were teachers themselves. Although he did not suffer fools gladly, Deming enjoyed responding to students' questions, even objections, about his philosophy.

Deming was a man available to everyone, regardless of position, an almost incredible feat in view of his manic schedule of seminars and consulting trips: he answered every letter, he honored requests for reviews of papers and books—usually quite promptly. He returned thanks for every gift, no matter how small. Here was the man who, after giving a four-day seminar in San Jose in January 1992, rode to visit our colleague Perry Gluckman on his deathbed. Deming, who often said, "We have not yet learned how to live," taught by his actions some effective lessons of how we must live. Some of us will not forget . . .

He often noted three ways in which one can influence others: through the power of office, by means of one's knowledge, and by the force of personality. Clearly Deming had no power of office: the movement he founded has no assets, no headquarters, no offices or office-holders; not even a list of members. His own business, unincorporated, had one employee, his assistant Cecilia Kilian. But Deming was a mighty influence through the other two avenues. On the morning of his death one disciple said simply, "I feel a light has gone out."

Deming typically started each of his seminars with a statement of aim: "We're here to have fun, to learn, to work together—and to make a difference." During the ensuing four days he would teach the principles upon which the prevailing style of interaction—among and toward co-workers, managers, subordinates, stockholders, creditors, suppliers, competitors, customers, the community—should be challenged and transformed. He called this set of principles a *system of profound knowledge,* taking little credit for its discovery and development.

This system forms a theoretical basis for the practice of management. It allows us to question most of what passes for management today—the so-called merit system, grading on the curve at school, "downsizing," the Baldrige award, to name but a few practices—and to know *why* it's wrong.

In 1974 the Nobel-prize-winning physicist Richard Feynman, in a commencement address at Caltech, made two observations that show much of what is wrong with management today, at the same time giving an unusual approach to a core part of Deming's philosophy.[1] We

---

[1] We are indebted to Dr. Arthur B. Robinson, President and Research Professor, Oregon Institute of Science and Medicine, for relating these details of Feynman's address to us.

might well spend a few minutes on his observations here. The first is that a *scientist's highest obligation is to prove himself wrong*. How paradoxical; yet this is equivalent to Deming's injunction to continually gain new and more useful knowledge using the scientific method. Management under what we call neo-Taylorism is management without examination of underlying theories, i.e., without opportunity to prove itself wrong.

Feynman's second point was closely related. He told of the natives of certain remote Pacific islands which furnished small but strategic bases for various Allied aircraft and military personnel during World War II. Often the arrival of one of these aircraft meant the distribution of many remarkable items to the islanders, gifts which made life easier and more enjoyable. Both the aircraft and the gifts were from a culture so advanced technologically that the natives were unable to comprehend them. But they knew they wanted the planes to keep coming, so they set out to discover what made them come there.

These primitive islanders had limited data about the arrival of the gift-bearing aircraft. They selectively analyzed their data to form a theory that correlated the arrival of the aircraft with airstrips featuring such unusual items as buildings, radio antennae, and so on. Evidence suggesting other causes of the visits, such as the need for fuel or to deposit supplies, was rejected. They pursued their theory by constructing mock airstrips, complete with "buildings" and "radio towers" made from local vegetation; islanders imitated air controllers by wearing "headphones" made of bamboo sticks. The planes came even more frequently. Until the end of the war the theory that the mock airstrips somehow caused the planes to visit seemed correct—the *form* of imitation was perfect, as Feynman observed—but not long afterward the flights ceased. No amount of "airstrip" operations would bring them back.

The cult leaders continued to reject all evidence that might be contrary to their original hypothesis. Had they admitted their theory was invalid, and tried to form a better one, the history of the "Cargo Cults"[2]

---

[2] The term "cargo cult" encompasses a number of remarkable cases dating back to the nineteenth century, involving the effects of South Pacific islanders reacting

would not be studied today. But cult leaders were about to receive some new evidence: they were given visits to modern civilization.

The aircraft and their operations on their islands were explained in some detail. Yet the leaders still rejected any evidence that contradicted their original hypothesis that the visits were caused by the crude artifacts they had created in the forest. They now had personal stakes in the cult they had created, and this new evidence clashed with their self-interest.

How closely this parallels modern American management! The United States dominated post-war industry, and the demand for American products was unprecedented: anything leaving our factories found buyers waiting to snap it up. Quality was an afterthought. Like the cargo cult leaders, American management committed the serious error in logic of assuming that mere correlation between two phenomena—prevalent management theory and soaring sales—proved that one caused the other. Top management adopted the theory that their success was caused by worshipping the quarterly P&L, by paper entrepreneurialism—endless shifting of assets among departments, divisions, and companies without creating any new value—and by adversarial relations with employees, customers, and suppliers. Government's theory was that it had contributed to success with its vast increase of taxes, tariffs, regulations, restrictions on trade, and political estrangement of key trading partners. But the success America enjoyed for three post-war decades occurred *in spite of* these actions, not because of them. And as with the cult leaders, their personal stakes in the current paradigm led them to dismiss evidence that directly contradicted their theory— from which they could have gained new knowledge.

Neither the cult leaders nor American management had much training in logic and the philosophy of science. They didn't know how to learn in the face of anomaly. At some point, however, innocence wears thin. After thirteen years of Deming's nation-wide teaching, western management typically still consider their job a blend of tech-

---

to contact with foreign cultures. Richard Feynman discusses cargo cult science in *Surely You're Joking, Mr. Feynman,* Bantam, 1989, pp. 308–317.

nique and reactive behavior. Their actions and concerns frequently led Deming to observe;

> *How would they know? How could they know? The answer is frightening: there's no chance to know. Without knowing what to do, we can be ruined by best efforts.*

It was against this ignorance—innocent, then cultivated and defended—that Deming fought the hardest and the longest. Something far greater than mere "quality control" is at issue here: Deming called it not just management but "the prevailing style of interaction" among all elements of society. Practising cargo cult science has kept western management—and the vast systems they control—in the pit of neo-Taylorism. By working harder, doing our best, embracing change, and so on, we only make that pit deeper, because the are not theory-based. We must know *what* to do!

Deming not only showed why neo-Taylorism is wrong; he also gave us a system of practical theory for transformation to a new style of interaction whereby everyone in the system wins—not just top management and Wall Street. With it we may know what to do, and by what method. This book is about that system of theory.

W. Edwards Deming's willingness to prove himself wrong allowed him to grow in knowledge, to be a good scientist. To us he seems to have made very few errors; one can imagine, however, that his mentor Walter Shewhart gave him many opportunities to correct these.

As we learn and pass along to others the sturdy edifice of theory he designed, we ought to remember that personal qualities such as intellectual honestly and willingness to learn are also essential elements of his system.

As eloquent as were Deming's words, his actions were probably even more effective agents of conviction. However you act, whether well or poorly, you *will* make a difference—in your family, your profession, your organization, and in society

One free man says frankly what he thinks and feels in the midst of thousands who by their actions and words maintain just the opposite. It might be supposed that the man who has frankly expressed his thought would remain isolated, yet in most cases it happens that all, or the majority, of the others have long thought and felt the same as he, only they have not expressed it. And what yesterday was the novel opinion of one man becomes today the general opinion of the majority. And as soon as this opinion is established, at once by imperceptible degrees but irresistibly, the conduct of mankind begins to alter.

LEO TOLSTOY

Introduction

1.1 When Will the Sleeping Giant Awaken?

Change has always been troublesome for people. In this book we are interested in two opposite kinds of change: chaotic change and smooth, controlled change.

Management, faced with the decline of their companies, typically look to causes outside themselves: the market, the competition, and especially the employees. In their attempt to wave a magic wand at disaster they push people against the wall, inciting factionalization and paralyzing them with fear. Under this pressure, uprisings and uncontrolled events will occur; some of these will represent progress and change for the good, but most will have effects that are unpredictable and often disguised. This is what we mean by *chaotic change*.

The transformation that Deming has been trying for years to encourage is one of *controlled change*. Controlled change is planned for by management and is designed for smooth transitions. When entropy, or disorder, is seen to be increasing, effort is applied to reverse the trend. We reject magic wands and silver bullets in favor of the wisdom of what we call *Deming's profound changes*. This kind of transformation is very difficult for many managers to accept because they are paralyzed in their current paradigm of service to self and to Wall Street. The result of such a transformation, however, would indeed be profound.

To avoid the chaos that could be ruinous to American society, the role of management is to lead the transformation of industry toward Deming's profound changes.

Just after his planes had bombed Pearl Harbor, Japan's brilliant naval strategist Admiral Isoroku Yamamoto, who had opposed war with America, said, "I fear we have awakened a sleeping giant." The Japanese, although their industry has advanced beyond ours in capability, still fear a day when the United States—which *they* consider a giant—will awaken, transforming itself according to Deming's principles.

What will awaken us? We hope that the rest of this book will not only sound the alarm but introduce the philosophy we feel is essential to that awakening.

Japan's Transformation From Taylor to Deming

Around 1980, W. Edwards Deming developed his list of Fourteen Obligations of Top Management. The list started with ten items, grew to fourteen, and has been improved many times since then. It has provided many thousands of his students with a rallying point in their attempts to spread his philosophy.[1]

About ten years later, Dr. Deming developed a new way to explain his philosophy: a system of profound knowledge. This system has offered the additional advantage of providing a base for further elaboration that was more difficult when starting from the Fourteen Points. Deming has urged others to help refine the system. Here is the best list of the elements of profound knowledge culled from Deming's writing that we have seen:

- Knowledge about the statistical concepts of variation
- Knowledge of the losses from tampering with a stable process, and missed opportunities for improvement of an unstable process
- Knowledge of procedures aimed at minimum economic loss from these mistakes (statistical process control)
- Knowledge about the interaction of forces (systems theory)
- Knowledge about losses due to demanding performance that lies beyond the capability of the system
- Knowledge about loss functions and problem prioritization (Taguchi loss function and the Pareto principle)
- Knowledge about the instability and loss that comes from

Notes for this chapter begin on page 41.

successive application of random forces (butterfly effect—chaos theory)

- Knowledge about the losses from competition for share of market (win–win versus win–lose)
- Knowledge about the theory of extreme values
- Knowledge about the statistical theory of failure
- Knowledge about the theory of knowledge
- Knowledge of psychology and intrinsic and extrinsic motivation
- Knowledge of learning and teaching styles
- Knowledge of the need for transformation to the new philosophy (management of change)
- Knowledge about the psychology of change[2]

In this book we use the concept of *profound change*. Profound changes are not a restatement of profound knowledge, but rather are its translation from theory to action. Thus, profound change is intimately bound to profound knowledge—the former is the result of applying the latter.

Although probably not following a preexisting plan, Deming chose to reveal his philosophy of profound knowledge to the world in two distinct stages:

1. **Profound Changes.** Taught to Japanese top managers in 1950, the profound changes concentrate on complexity, variation, and the use of the scientific method as a tool for learning and improvement. Deming knew that Japan was under the sway of Frederick Taylor's theory of scientific management, as was the West; he knew they would not make any progress until better theory was provided. Additionally, Deming revealed to Japanese top management that the systems they must improve and serve include elements traditionally considered "outside" the system, such as customers, suppliers, and even competitors.

2. **Axioms.** Taught to Western managers and students beginning in the 1980s, these comprise a kind of unified field theory for whole-system improvement reflecting Deming's system of profound knowledge.

Our presentation in this book follows the same progression, from profound changes to axioms. It is better, for example, that people—especially managers—learn first about reducing complexity in their operations, and then later find out why they ought to abolish management by objective (MBO) and the merit system. Such a sequence may help many to understand and accept Deming's philosophy who might otherwise challenge it.

2.1 Deming's Profound Changes

In 1950, W. Edwards Deming, a U.S. statistician prominent in the field of sampling who had been a leader in the wartime effort to educate engineers in statistics for quality control, went to Japan to teach classes on a method of management that he had helped Walter Shewhart (1891–1967), a physicist at Bell Telephone Laboratories, develop between 1935 and 1940. As a result of these classes, most of the large Japanese companies—such as Toyota, Mitsubishi, and Matsushita, whose names have become household words in America today—made momentous changes in the way they were managed. We call these Deming's profound changes, and will enumerate them in a later chapter.

To understand the enormous effect that Deming's profound changes had on the productivity increase of the Japanese, it is necessary to understand not only Deming and Shewhart's school of management but also that of the American engineer Frederick Taylor, the father of Scientific Management. The validity of Taylorism and its extensions is an unstated assumption in almost all U.S. companies today, making it difficult to discuss any other mode of running a business.

The relationship between Taylor's school and Deming's school is similar to that between the Taylor school and the craft system which it so effectively displaced. The differences and similarities between Taylorism and Deming's philosophy are not, however, as easy to understand because the implementation of Taylor's philosophy has changed with today's world. Understanding Deming's profound changes is a good place to start.[3] By starting with the profound changes, in a short time any company willing to make these fundamental changes can become a top competitor, increasing their productivity and profitability and financing their own growth.

Deming, who had been already widely known for his work in sampling theory, taught sampling and statistical control charts for the War Production Board during World War II. The teaching efforts were shared with such colleagues as Harold Dodge, Eugene Grant, and Holbrook Working. Their chief audience was industrial and governmental quality inspectors, who worked with the untrained work force in the factories to use Deming and Shewhart's principles, teaching them how to use statistics to control quality. More than ten thousand were trained in these wartime programs. Although by no means spreading to *all* U.S. industry, the increases in productivity and quality of war matériel and other goods that followed from this teaching[4] were as if a giant were waking up—the analogy used by Admiral Yamamoto. Deming later built on this technical base when he taught Japanese top management beginning in July 1950.

Deming and Quality

The philosophy of Deming outlined in this book is qualitatively different from what others are teaching in the quality field. An example of this difference is the statistical control chart. The control chart, invented by Walter Shewhart in 1924, offers a method of determining if the cause of variation in quality is a *common cause* occurring within the process, or a *special cause* occurring outside of the process but affecting its outcome. In order to improve in the most effective way, an organization must recognize the predominant cause of variation in the *see note 5.*

process and devise an approach to find the cause based on observations. Without this kind of knowledge based on data, tampering will likely occur, resulting in increased variation rather than improvement.[5]

AC's
systematic us random error

Shewhart's main purpose for the control chart was to distinguish between the two causes of variation and help management avoid making mistakes in dealing with them. If a process has special-cause variation, it first has to be stabilized by removing the special causes. For stable processes the task is to reduce the common-cause variation from within the process. This will increase the ability to predict a tighter, higher-quality result from the process—variation will decrease and the economic loss from the process will drop. Shewhart proposed the use of the scientific method (Deming today calls this the Shewhart, or PDSA, cycle[6]) to gain the knowledge about the process that is needed in order to isolate and remove common causes.

cf Type I and Type II error

cf signal vs noise

Over the years, however, control charting has in many cases become a technique for merely keeping the process in control (so that the only variation is common-cause variation), and no attack is mounted on the common-cause variation itself. This has often been described as statistical process control. In worse cases the data, which represent the voice of the process, have been distorted in order to make them fit a mathematical formula or distribution—which is controlling the chart rather than the process.

cf SDT

As a result, Shewhart's highly effective *improvement* technique of looking for the predominant form of variation and finding the flaws has been considerably weakened when reapplied in the United States. The failures which occurred when many of the modern improvements were copied from Japan were, as you'll see later in this book, the result of what is called extension transference—a dangerous form of imitating but missing the point.[7]

φ ?

As Deming recognized, Walter Shewhart had developed a method of understanding the nature of variation, providing management with the ability to respond appropriately. In essence, the path toward continuous improvement is in learning how to listen to the process and the system (collecting empirical data, rather than making changes intuitively) and in using the scientific method to isolate flaws and remove them from the process or the system.[8]

A common error concerning W. Edwards Deming is to assume that his philosophy is limited to product and service quality improvement (as is that of most of the other quality experts) or that he is "the prophet of total quality." We want to get Deming off the quality shelf, and have him recognized for his contributions to a unique philosophy of life, a world- and life-view that applies to parents, teachers, and other members of communities, in addition to the typical management audience. It is not lightly that we employ the term "profound" to what Deming has done. In *Deming's Profound Changes*, we provide clear, helpful insights which can be used to learn and apply the Deming philosophy.

2.2 Japan's Important Change

In 1950, the majority of Japanese industry changed from Frederick Taylor's school of scientific management to a school that was taught to them by W. Edwards Deming. For the Japanese, who had been in the throes of Taylorism for almost half a century, this was a major change. They were willing to make the change for at least four reasons:

1. **Deming told them if they made these changes they could capture the markets of the world.** Japan, a nation virtually without natural resources, had embarked on Southeast Asian adventurism as a way of assuring a supply of resources, finally resorting to an irrational war with the West. Both strategies had failed abjectly. Deming now invited them to focus on demand instead of supply by using *quality and grade of product as a national strategy* for the gaining of security and national advantage through production and world trade.

2. **They believed Deming's teaching during the war had an impact on the outcome of the war.** The Japanese had estimated the U.S. would only be able to produce a certain

amount of war materials in nine years to fight a two-front war. Instead, the United States, with Deming's methods playing a role, was able to produce twice that estimate in three and a half years—using a less experienced work force than prior to the war ("Rosie the Riveter") and without all workers being mobilized towards the production of war materials. One reason Deming's teaching gained easy acceptance during that time was because much of the experienced management and work force—accustomed to traditional, Taylorized thinking—was being used for combat.

3. **The changes Deming suggested agreed with the Japanese culture to a greater extent than did the Taylor school.** Japanese culture recognizes that many small steps toward total system improvement are very acceptable, can come from all areas, and can benefit everyone. Deming's view of production as a system, rather than a hierarchy of independent units, was congenial to their views of business and society. Further, the Japanese find it very natural to work together in teams to find flaws in systems. Western culture typically optimizes only parts of the system and exalts the contributions of visible individuals, at the expense of the total system. Long-term, holistic thinking, as used in Deming's philosophy, is inherent in the Eastern culture.[9]

4. **Japanese top management were absolutely committed to improvement.** The Japanese paradigm shift was clearly driven from the top down. Dr. Deming strongly contends that the United States will be unable to adopt the new philosophy until top management leads the change.

As an example of how productive the United States had been with Deming's changes, one speaker described how Boeing had built one B-17 bomber every forty-five minutes; according to another speaker, a battleship could be built in five days. The difference between Taylorism and what Deming added was simple: continuous quality improvement using statistical principles. If the U.S. had continued to use the changes Deming taught, it would have no problem competing with the Japanese for world markets today.

One important way American industry stopped using Deming's profound changes was in sending Rosie the Riveter back to the kitchen when the experienced work force returned from the war; the professional management team reverted to the Tayloristic practices of the pre-war period, as with the methods of Alfred Sloan at General Motors. Although a few business schools have recently begun to give some sway to Deming's ideas, Taylorism still pervades most business school curricula.

If the U.S. is going to increase its productivity it will have to make the same profound changes that the Japanese made in 1950, *and more.* Managers will have to be trained in Deming's principles instead of Taylor's. The rest of this book details and explains those principles, starting with Taylor's.

In the course of their long history the Japanese have become a survivor nation, not only resisting the attempts of China to invade and dominate them but also adapting to outsiders from the West such as Commodore Perry in the mid-nineteenth century. They survive by emulating, then improving upon, the secrets of their adversaries' strength; and then by beating them at their own game. Japanese culture, religions, and even language are for the most part imports. Instead of "not invented here," the Japanese attitude seems to be "Anything you can do, we can do better—without losing our own essential nature and character."[10] Thus, even if Deming's profound changes had not proven as congenial to its culture as turned out to be the case, Japan would very likely have found some ways to adopt its essence, make it uniquely Japanese, and use it to advantage.

> *Will the United States have to be in ruins before Deming's profound changes are accepted here?*
>
> *Will most of our property and productive capacity have to be in the hands of foreign owners before we will begin to learn from our mistakes?*

Will managers who learned how to make the system serve them
continue to direct our course?
Is the United States a survivor?

A Japanese Commentary on Taylor

We will win, and you will lose.
You cannot do anything about it because your
failure is an internal disease.
Your companies are based on Taylor's principles.
Worse, your heads are Taylorized, too.

Konosuke Matsushita
Founder, Matsushita Electronics
1988

Japanese leaders realize that as long as the U.S. does not change from the Taylor school, Japan has an insurmountable advantage over the West. These leaders also realize that the United States is a sleeping giant—one who could awake at any time, just as happened during World War II, putting their nation once again at a serious disadvantage. They certainly still watch the U.S. to see if we will drop our current practices and adopt an approach that emphasizes quality with the success the Japanese have enjoyed.

This is, however, not a book about Japan or Japanese management. Before we shift our focus to the West, it would be well to ponder the paradoxical fact that, although tradition typically resists change, change itself is part of the Japanese tradition. We do not know what forms and arrangements Japanese business will take in the future, but we do know that Deming has often criticized their management for imitating Western practices.[11] To the extent that this imitation of the West increases, Matsushita's prediction of Japan's victory will be less and less certain.

U.S. managers think that Taylor's philosophy fell out of use a long time ago. Actually, much of what is taught in business schools today is still based on Taylor.[12] The best way to see this is to compare what is being taught in the schools today to the axioms contrasting Taylor's and Deming's philosophies which are presented later in this book.

Konosuke Matsushita is right: our heads are Taylorized too. In other words, Taylor's philosophy is shown both in how things are done in American companies as well as in how American managers think.

Changing from Taylor's axioms to Deming's is the prerequisite for America's recovery.

2.3 Taylorism and Neo-Taylorism

What is Taylorism?

The term *paradigm* formerly meant simply an example or pattern (useful in the teaching of mathematics), but today it has come to mean, more broadly, a system of rules and practices that allow people to function within that system. Paradigms tell us how to act to be successful within them, and what to expect from other people. A paradigm can be formal—as in the military system with its ranks, uniforms, and regulations; or a paradigm can be informal—as in the customs, dress, and communications by which business is generally conducted. In Western business, Taylorism and its modern manifestation neo-Taylorism are the dominant paradigm of management today. Thus, understanding Taylorism is essential if Deming's profound changes are to be appreciated and enacted.

Who was the man of whom Peter Drucker would say, "I do not think it extravagant to consider Frederick Taylor as the one relevant social philosopher of this, our industrial civilization"?[13] What are Taylorism and neo-Taylorism, the systems based on Taylor's philosophy? Without the answers to these questions we are like fish who swim and live their lives in water yet don't know of water's existence. Please join us for a short, but vital, trip into the past.

Frederick Winslow Taylor (1856–1915) was an American engineer who formulated a theory of scientific management, which was the culmination of scientific and managerial thinking back through the mid-eighteenth century. Scientific management and the resulting industrial efficiency movement, which became famous for its use of time-and-motion studies, were direct attempts to abolish the craft guild system, which had dominated shop floors for centuries, by applying a deterministic philosophy to business operations.

Determinism argues that the future (of people, things, systems) is controlled, or *determined* entirely by history, ruling out novel and unpredictable events that may arise out of interactions among causes. Determinism is usually accompanied by a belief that we are well along the road to knowing history sufficiently to make such predictions with complete accuracy.

The framework of scientific management was production quotas enforced by new pay and personnel systems, designed to require workers to meet scientifically determined work standards which were well above the then-accepted norms. Workers who accepted the scheme typically doubled or tripled their previous output—and increased their wages by 60%. As Matsushita-san would say almost a century later, management was assumed to have the good ideas and the workers' role was to carry them out.

Scientific management was highly successful at increasing the industrial output of the United States.[14] And, it was quite appropriate for the times, given the high influx of immigrant workers, along with the resulting mix of languages and cultures. Yet in the light of the modern management theory of Shewhart and Deming, Taylor's system was substantially flawed.

Worse, Taylor would probably recoil in horror at what has become of his original philosophy today. We will deal with this outcome later in this chapter as "Neo-Taylorism."

Taylor's Forebears

As we will see, a hallmark of scientific management is Taylor's mechanistic view of both the organization and the worker. Taylor was far from the first to hold such a view. One of the first men to write about business—as we would know it today—and its problems was Josiah Wedgwood (1730–1795), from whom Wedgwood china takes its name. Frequently at odds with his workers over compensation, Wedgwood correctly identified mass production as a way of producing the most in a given period of time. When his company went through financial distress in 1772 he studied costs at each stage of his processes. Most importantly for our purposes here, Wedgwood may have been the first to want to "make such machines of men as cannot err."

A generation later, the social reformer Robert Owen (1771–1858), known for progressive treatment of workers at his textile plant in New Lanark, Scotland, nevertheless considered them "living instruments" and "vital machines."[15] We will return to Owen in another context much later in this book.

Charles Babbage (1792–1871), one of the fathers of the computer, opined that management was a science, not an art, and restated the economist Adam Smith's arguments for division of labor. Unhappily, in his work *On the Economy of Machinery and Manufacture*[16] he offered cost control and empiricism as that science. Both Henry Metcalfe (1847– 1917) and Henry Towne (1844–1924) favored the application of science to improve processes by means of recorded data about them.

Four themes seem to develop in the school of thought these men represented.

1. Businesses have existence separate from their owners.

2. The purpose of a business is to make a profit for its owners.

3. The worker is like a bionic machine, in whom loyalty, imagination, or enthusiasm are considered at best neutral qualities—at worst, liabilities.

4. Scientific and engineering methods should be used to improve business processes.

On the scientific side, Taylor's chief forebear is much better known: Sir Isaac Newton. Newton's very tenets of mechanism and predictability were what these early commentators wanted to apply to business. We will return to Newtonian thought later.

The Birth of Scientific Management

The management philosophy of our age is captured in the almost-synonymous terms "Taylorism" and "scientific management." The son of a successful lawyer in Philadelphia, Taylor attended the Phillips Exeter Academy[17] and intended to enter Harvard University, but was prevented by eye problems said to be caused by night study. Instead, in 1874 he entered industry as an apprentice machinist at the Enterprise Hydraulic Works in Philadelphia, while continuing the activities typical of his upper-class status, such as playing tennis and cricket. Taylor completed his apprenticeship in 1878 only to be unable to find work in his trade, during the depression following the panic of 1873. He joined Midvale Steel Corporation as an unskilled laborer, rising eventually to a position in management and earning a degree in mechanical engineering from the Stevens Institute of Technology along the way.

Frederick Taylor had developed a reputation early in life for being different. He exasperated his early playmates by insisting on elaborate and strict rules for each game they played. At the age of 23 he was a member of the national tennis doubles championship team, and is claimed by at least one author to have been the first man in America to throw a baseball overhanded. Taylor held over forty patents, and in his last years invented a two-handled putter which was quickly banned from the links by the U.S. Golf Association. In anything he pursued, Taylor exhibited a frenzy for order, discipline, and optimization.

At Midvale he held a variety of jobs which allowed him to gradually develop and apply his theory of scientific organization of human work to achieve optimal processes. Biographers call him a man of two personalities: the ambitious son of genteel Quaker upbringing ("thee" and "thou" were still spoken at home), and the upper-class factory revolutionary using his studiously cultivated vocabulary of salty language to bully the men into following his new standards of efficiency.

Soldiering and Pay Incentives

Taylor's career as an unskilled laborer soon brought him into contact with deliberate output restrictions by the men, which he later called *soldiering.* Taylor classified two types: *natural* soldiering ("from the natural instinct and tendency of men to take it easy") and *systematic* soldiering (from the workers' careful thought and reasoning toward what they felt would best promote their interests). Systematic soldiering had, he said, "the deliberate object of keeping their employers ignorant of how fast work can be done." Taylor ascribed the continuation of soldiering to management's ignorance of what output levels were actually possible.[18]

Soon after he was promoted to gang boss, Taylor reported, he was approached by one of his former mates who appealed to his sense of fellowship, warning him that if he tried to get the team to increase their work rates they would throw him over the fence. Taylor ignored such threats and, over the next few years, struggled to create the system of shop management that now bears his name.

One reason for soldiering, Taylor came to realize, was the failure of existing pay schemes to provide real incentive to raise production. Employers tended to reduce piece-rates, for example, when worker output even briefly increased, which taught the workers to collaborate in bringing their productivity back down. He admitted that, "if he were in their place he [too] would fight against turning out any more work, just as they were doing, because under the piece-work system they would be allowed to earn no more wages than they had been earning, and yet they would be made to work harder."[19] Taylor's attempts to replace this antagonism with "harmonious cooperation," through production-based incentives and innovative piece-rate schemes, were to form one cornerstone of scientific management.

Another key observation of the same period was that management were ignorant of the details of the tasks to be done in their plants. Taylor saw that skilled workers were organized under the craft guild system with its progression from apprentice through journeyman to master of a particular trade. Under this scheme, the master craftsmen ran their respective shops as though they were managers, controlling not only production, but training and, of course, promotion to higher ranks. Although he held the interests of worker and management to be complementary, Taylor considered the control of the shops by the master craftsmen both inappropriate and inefficient, and he resolved to place control in the hands of management where he felt it belonged.

de-skilling

Three themes are clear in Taylor's thinking of this period:

1. The concept of business as a vast and complex mechanism amenable to scientific analysis and, ultimately, control and prediction in Newtonian terms via standardization of all work

2. The inappropriateness of the craft guild system under "ordinary management," which kept the know-how of production in the heads of the workers and foremen

3. The unequal division of labor between management and worker

These beliefs about the current system, combined with his mania for efficiency, led Taylor to begin experimenting when he rose to supervisory level. He broke tasks down into the smallest movements, and then determined what the "optimum" time was to complete each movement, how much weight was to be carried and how far, when the worker should rest and how long, and so on. If a task could be done by a less-skilled worker, it was shifted away from the skilled one. From these experiments the term "time-and-motion study" originated.

Only the Best Workers

After these studies Taylor would set work standards and quotas that even "first-class men" would have to strain to reach. Taylor's protes-

to quote Fred

tations of scientific accuracy notwithstanding, work standards derived from his system were substantially influenced by subjective considerations, such as defining what activities actually constituted the job, which workmen were to be measured, and what allowances were to be made for differences in materials and uncontrollable interruptions.[20] He also studied how much additional pay would have to be offered to get the men to achieve the new levels, and settled on three pay groups ranging from 30% to 100% premium over what was paid for "average work." Not surprisingly he now called his methods the "task system"; for the workman, its most prominent feature was "the setting of a measured standard of work for each man to do each day," for which he was well paid—if he could meet the work standards. If not, he was severely penalized by Taylor's new piece-rate pay system.[21]

Despite initial resistance by some who had not been selected to be on piece-rates, the scheme caught on quickly, with men waiting to be put on the new task system. Taylor extended its application from the shop floor to administrative and sales areas.

The result was a far more regimented workplace controlled by management according to Taylor's "one best way of doing things." Surprisingly, though, Taylor expected a great deal more, not less, from management: according to his new scheme, management should do about 50% of the total work of the organization in its activities of selecting, training, and motivating workers. Supervisors were expected to work closely and continuously with the men to see that the new standards were adhered to. Where this philosophy was followed, productivity rose.

Although he was later to be vilified by organized labor, Taylor's early experiments were largely popular with the men, as he would argue to management that they should receive raises proportional to their increased marginal productivity. His belief in optimal systems and management control soon outweighed his sympathy for the men, however, and he became a champion of work standards and pay systems that forced the worker to follow his standards and be first-class men, or be replaced. Taylor clearly intended to run the place with only the best workers. He rationalized this to himself on the basis that scientific management always created more jobs, which the ordinary men who had been forced out could take.

Success and Controversy

In 1898, Taylor accepted an offer from Bethlehem Steel Company to come and work exclusively on the application of his system. He also found time to do research there with the metallurgist Maunsel White, which resulted in the development of a formula that revolutionized the machining industry. Their tool-steel alloy, heat treatment, and procedures for using the tools allowed tools to cut metal at rates several times faster than possible before. Their work was recognized with the award of a gold medal at the Paris Exposition in 1900.

In 1901, at the height of his career, Taylor left Bethlehem over a disagreement with management: he had demanded that the men be given more pay and time for leisure in proportion to the increased efficiency his system brought to their work. Until his death, he offered his consulting services free to anyone interested in "the one right way to do a thing"; his business card read "Systematizing Shop Management and Manufacturing Costs a Specialty." For the remainder of his life, however, Taylor was relatively inactive, even turning down an offer to become president of the Massachusetts Institute of Technology.

The effects of Taylor's theories on turn-of-the-century American industry were nevertheless rapid and substantial. The manufacturing world that Taylor entered had been closer to craft production than to manufacturing as we know it today; by the time of his death in 1915, the improvement of production under scientific management formed a logical extension which permanently changed business from its haphazard earlier state. The mass-production era under modern management had begun.

Of course Taylor was not the only laborer in the vineyard. Men like Sorensen and Flanders, and their boss Henry Ford, spent six years experimenting with moving assembly lines before installing the now-famous one at Highland Park in about 1913, and their concerns were much broader than Taylor's. At Ford, the goal of 100% interchangeability of parts meant that having a man do a job was a temporary last resort if no machine were available yet to do it. In addition Ford achieved significantly greater division and concurrency of labor in his continuous assembly lines than Taylor ever did working with individual laborers. But "Speedy" Taylor's paradigm of efficiency and standardization was already a part of the spirit of the automobile age, and at least part of Ford's work can be seen as an extension of Taylor's.

Not all of the consequences of Taylor's ideas were necessarily to his liking. By 1911, when his book *The Principles of Scientific Management*[22] was published, Taylor's methods were the subject of national debate, and he himself had become the darling of various movements that he probably despised. That same year, pressure from trade unions and other parties caused the U.S. House of Representatives to convene a special committee to investigate Taylor's methods; Taylor himself gave lengthy testimony. In addition, Lenin called for adoption of scientific management in the organization of Russian factories under Communist rule.[23] For better or worse, Taylor's management philosophy, with inevitable corruption, has dominated the twentieth century, the one major exception being post-1950 Japan.

Scientific Management Principles

In his book, *The Principles of Scientific Management* (1911), Taylor listed four principles, shown in the following list. We have elaborated on the implications of each principle:

1. Develop a science to replace the old rule-of-thumb knowledge of the workmen, which the workmen kept in their heads. Reduce this knowledge to laws and rules and formulae.
   - Assumes a Newtonian clockwork universe

- Provides no principle of learning, just a transfer of knowledge (no indication of *how*)
- Suboptimizes at the task level
- Expects identical results in future
- Does not consider variation
- Breaks the craft guild system
- Gives revolutionary, discontinuous improvement

2. Scientifically select the workmen, then progressively develop them.
   - Makes Newtonian assumptions of knowledge and ability to select the workmen

3. Bring together the science and the scientifically selected workmen.
   - Provides extrinsic motivation
   - Conditions company's loyalty to both workers and management on compliance with work standards
   - Demands quotas for above-average performance from all workers

4. Divide almost equally the work of the establishment between the workmen on one hand, and the management on the other.
   - Assumes that everyone needs to be supervised
   - Creates multiple layers of supervision to enforce standards and quotas
   - Creates ladder for ambitious managers to climb

Taylor's Disciples

When Taylor died in 1915 his theories continued to be preached, fairly intact, by his disciples. One, Carl Barth, had joined Taylor at Bethlehem and had become an unquestioning adherent of Taylor's system, referring to his practices as "our method." Barth became known for two things: a slide rule which made Taylor's empirically determined metal-cutting relationships (true contributions for which Taylor is still recognized today in metallurgical circles) rapidly accessible to the engineer; and for verbal abuse, with which Barth would quickly incur the hostility of clients.

The next two disciples are fairly commonly known by name even today. Henry L. Gantt (1861–1919), whose name is attached to the familiar planning chart, had worked with Taylor at Midvale, and wrote three books and several papers giving the standard Taylor line. His description of scientific method was closer than Taylor's to what we would know today, but he, as did Taylor, considered operations only individually without reference to the system of which they were parts. Gantt, well-liked for his respect for and accommodation of client management, favored paying foremen bonuses whenever their teams met their work quotas, claiming this motivated the foreman to "give special attention to the men most likely to fall behind." Gantt was even more explicit than Taylor about what the workman's job was ("to obey orders"), although he considered the boss to be the servant of the men—but only "so long as the workmen perform their tasks."

Frank B. Gilbreth (1868–1924) is perhaps the only Taylorist to have appeared, although posthumously, in the movies. He, his wife Lillian, and their large family were the subjects of the 1950 film *Cheaper by the Dozen*, which provoked great hilarity by showing the results of the parents' obsession with getting the family to function "the one best way" using time-and-motion studies. Gilbreth wrote at least seven books on scientific management, and extended Taylor's concepts to the work of managers, office workers, and even the handicapped. His fanatical approach is evident in the statement, "The super-standardization of everything, every tool, every practice, every process, down to the most insignificant items, is necessary for the greatest efficiency."[24]

Not all the writers of the period toed the orthodox Taylor line. Mary Parker Follett (1868–1933), author of several books on management, carried on some of Taylor's themes but formulated some concepts that show significant breaks with scientific management, such as in her description of leadership qualities: "The leader must understand the situation, must see it as a whole, must see the inter-relation of all the parts."[25] She also espoused the systems viewpoint (which she called "integration"), cross-functional teamwork, and collective responsibility.

Even within the Taylor Society, the following advice to management, which sounds very much like Deming's own, was published in 1922:

> *Profits and losses . . . are a final result; by the time we get them it is too late [to take action]. What we want is information on the things from which profit and loss result while there still exists time to affect them.*[26]

But such are exceptions; Taylorism was the rule.

Neo-Taylorism

We've noted that scientific management succeeded in its time. Taylor's system helped transform American industry from craft production to mass production, and mightily influenced its management philosophy. Given the state of knowledge available to Taylor, his ideas and methods represented a logical extension to previous modes of thought, and permanently affected not only industry but management practice everywhere, including Japan. Only decades later would there be any widespread understanding of some of what we now know to be serious, basic flaws in scientific management; we will see this in the discussion of modern physics ("Philosophical Basis of Taylorism," later in this chapter), as well as in the work of statistician Walter Shewhart, Deming's mentor, in the late 1920s.

No one will probably be shocked to learn, however, that Taylorism and scientific management are not mentioned much in business these days, nor in business schools. What happened between all that success in the early part of the century, and now, when American business troubles are on the cover of even non-business magazines?

Two things happened:

1. Scientific management became outmoded by the progress of science and philosophy.
2. As with many ideas which are good for their time, Taylorism was corrupted over a long period by both a gradual shedding of its good aspects and a continuing admixture of new management philosophies that required little more of management beyond hiring, firing, and the giving of orders—and maintaining an image.

Taylorism became neo- (or "new") Taylorism. It was not a deliberate transformation; no one said, "Let's modify Taylor's principles so that we can work less at them." Citing an exact period for this transformation would be difficult, but most of the features of neo-Taylorism seem organic in the management philosophy that has been with us in the West since the end of World War II.

Flaws of Scientific Management

What, then, are the flaws of scientific management that we speak of? They are many. For each of the flaws shown below we have included

1. Belief in management control as the essential precondition for increasing productivity
2. Belief in the possibility of optimal processes
3. A narrow view of process improvement
4. Low-level suboptimization instead of holistic, total-system improvement
5. Recognition of only one cause of defects: people
6. Separation of planning and doing
7. Failure to recognize systems and communities in the organization
8. View of workers as interchangeable, bionic machines

Table 1. Flaws of Scientific Management

[handwritten margin notes: "this is a huge hand-wave. What are the specifics here? I do not know"]

[handwritten margin notes: "These are only 'Flaws' re' arbitrary' goals to which Taylor(ism) + Neo-Tâism would probably not subscribe, so they've only Flaws if you accept the (unstated) goals of the Deming crowd"]

[handwritten margin notes: "cf Reason re latent design error"]

a commentary to show its effects in our own time under the heading of neo-Taylorism.

1. Belief in Management Control as the Essential Precondition for Increasing Productivity

- Formerly, shop-floor workers had the monopoly of production knowledge; improving productivity meant management had to wrest that control from them.
- Workers were assumed to routinely restrict their potential output, and to conceal their knowledge of how to increase production.
- Management believed in enforced work standards and even enforced cooperation.

Taylorism in Action Today (Neo-Taylorism)

- Your boss is your customer.
- Management increasingly answer to stock market analysts and bankers.
- Management are preoccupied with manipulation of the work force, especially through the dual means of reward and punishment.
- Management are obsessed with control and results, rather than improvement. *isn't this just a duality?*
- Planning is reserved as a managerial activity.

Contrast Deming's insistence that management's job is not to control but to lead, to coach, to provide methods and tools. The result is control—statistical control. *i.e., don't argue w/ the "laws of physics"*

2. Belief in the Possibility of Optimal Processes

- Taylor's deterministic outlook assumes static laws and relationships govern the performance of any system, hence all business systems; the inherent randomness in all processes is unrecognized. *DDW's unpredictable variability*
- Process optimality (Taylor's "one right way to do a thing") assumes the absence of variation; so variation in a worker's

satisfy + soffice = satisfice

as opposed to many ways to satisficingly solve a problem

output was ascribed to soldiering or lack of desire to be a first-class man.

- All needed information is assumed to be knowable.
- Everyone was required to produce at levels well above those of even first-class men, enforced by a pay system that heavily penalized substandard performance.

Taylorism in Action Today (Neo-Taylorism)

- Perfectionism—a belief that "best efforts" will achieve optimal results
- Management by objective (MBO) as a vehicle for enforcing perfectionism
- Rejection of continuous improvement *[handwritten: where has this happened, at least in lip service?]*
- Belief in certification as a guarantee of quality
- Demands for repeatability of processes
- Denial or minimization of failures, and dismissal of the opportunity to learn from them
- Reliance on reengineering and automation as a substitute for reducing inherent process complexity
- Spreadsheet mentality in planning (assumption that fixed or linear relationships exist among variables)
- Quotas and work standards
- Hiring only high-GPA college graduates
- Employee ranking and rating schemes
- Imitation of others without understanding why, how, or even *if* they get their results
- Multiple suppliers for the same product when a single-supplier relationship would be feasible
- Concentration on measurement of outcome rather than understanding of the underlying system of causes

[handwritten margin notes: Show me →; show me →; hyperbole?; à la out-comes-based education]

Contrast Deming's emphasis on continual learning and improvement, and the unknowability of the most important optimization data.

Contrast Deming's emphasis on understanding of variation, special and common causes, and statistical control; also, his pleas to scrap the "merit" system.   *all measures are operationally defined*

Contrast Deming's and Shewart's position that there are no true values (of observed phenomena) against management beliefs that everything is quantifiable and knowable.

Contrast Deming's insistence on using the Shewhart Cycle for never-ending learning in order to understand and reduce variation.

3. A Narrow View of Process Improvement   *suppose it were — how defined in any case?*

- Minimal, if any, focus was on product quality.
- Improvement was limited to selecting, training, motivating, and supervising the workers (these are in fact the areas addressed by Taylor's four principles of scientific   *on p.20* management).
- An "optimum" process was selectively determined off-line by experts, meaning people trained in scientific management.
- De-skilling of tasks (reducing the number of functions included) to reduce costs and enable time-study often left skilled workers "over-qualified" and therefore "over-paid"; this move was partially aimed at the disfranchisement of   *27* skilled labor.[27]
- Time-studies sometimes used idealized and subjective data taken from other processes. ("Mere statistics as to the time which a man takes to do a piece of work do not constitute 'time study.' 'Time study,' as its name implies, involves a careful study of the time in which work *ought* to be done."   *woof!* [Taylor])[28]   *28*
- Concentration was on minimums (time) and maximums (output) rather than on averages.
- Introduction of new processes and techniques developed off-   *? to what does this refer?* line was occasional and discontinuous.

Taylorism in Action Today (Neo-Taylorism)

- Manipulation of the *symbols* of progress, instead of real improvement (looking good vs. being good). (See section 4.2, "Phases of Learning.")

- Chasing awards, prizes, and publicity instead of pursuing improvement
- Management reorganization used as a substitute for process improvement
- Adoption of spurious measurement systems which ignore variation
- Adoption of current fad programs as "the one best way"
- Satisfaction with merely meeting specifications
- Reliance on technology, planned invention, and "silver bullets" instead of on continuous improvement

probably means "not "zingers" silver bullets"

- Emphasis on documentation rather than on improvement of processes *(but documentation (eg of design rationale) can improve a process)*
- Lack of emphasis on continuous learning, and an unwillingness to learn from the past
- Teaching immediate job skills instead of educating people to think, analyze, and gain new knowledge

Contrast Deming's use of scientific method (PDSA cycle) to isolate and reduce the flaws and complexity causing variation.

Contrast Deming's (and others') principle that any system is the product of the interactions among its parts, not the sum of individual performances of the parts.

Consider Deming's position that, as powerful as process improvement is, the *truly* great leverage in improving an organization comes from treating problems of the system, e.g., how people are regarded and treated ("We haven't yet learned how to live.").

Contrast the ideal of developing and increasing knowledge as described in Ikujiro Nonaka's paper "The Knowledge-Creating Company" (*Harvard Business Review*, November–December 1991).

4. Low-Level Suboptimization Instead of Holistic, Total-System Improvement

- Each person's work was redesigned for maximum output, and work standards are based on that maximum.
- No concept of continuous improvement existed—Taylorism always aimed at "the one best way."
- Variation was taken out of the equation, and the mean (or some other arbitrary figure) was made deterministic (repeatable every time given the same initial conditions) for the process.
- Each unit of production was "made best" individually, not holistically as part of a system.

Taylorism in Action Today (Neo-Taylorism)

- Suboptimization of one component at the expense of another (win–lose), for example, "Run your department like a business."[29]
- Quotas, goals, and targets for various groups, which increase complexity through competition, factionalization, and duplication of effort, instead of benefiting of the whole system:

   Sales, shipment, and inventory quotas
   Zero-defects targets
   Cost-reduction goals
   Arbitrary staffing-level constraints
- Incessant reorganizations to meet some "new challenge"
- Intolerance of "idleness" on the part of workers or machinery, resulting in:

   Production bottlenecks
   People running out of work from time to time, and keeping busy by inventing new projects
   Engineers developing several versions of the same product at one time

Make-work projects caused by failure to allocate resources
to the bottleneck

Build-up of work-in-process inventory

Lengthening cycle times

Rising production costs

- Bottlenecks shifting around after great (and often expensive)
effort to eliminate them

- Measuring the efficiency of subgroups and individuals
(productivity indices, units per hour, transactions per hour),
often resulting in factionalization and competition to make
the measurements better

Wow

Contrast Deming's analogy of a symphony orchestra, whose peak
performance does not depend on maximizing the performance of each
individual unit. When *each component* of a system maximizes its own
performance, the *system as a whole* will not perform as well as possible.

Contrast Deming's insistence on solutions where "everybody wins."

Contrast Deming's emphasis that the randomness and complexity
inherent in all processes makes the most important information about
them unknowable.

7
a

Contrast constraint theory's emphasis on balanced *flow* rather than
balanced *capacity*.

5. Recognition of Only One Cause of Defects: People

and by "people"
the authors mean
worker - people,
not mgmt - people
or designer -
people.

- "Optimal" systems don't produce defects; hence, the only
variable must originate outside the system: the worker.
("Lessons of the Red Bead Demonstration," section 6.7, is an
excellent case study!)

- Workers were assumed to deliberately restrict output in order
to prevent management from learning their true capacities.

- As Taylor stated, "Scientific management is a complete
mental revolution on the part of the *working man*."
(Emphasis added.)

- Pay schemes, tied to arbitrary speed and production quotas,
were designed to "convince" workers that their interests and

management's were common (namely, meeting the quotas).
Taylorism in Action Today (Neo-Taylorism)

- Worker-motivation schemes, such as:

    "Employee of the Month" and similar awards

    Making employees sign pledges to do quality work

    Slogans, posters, films, and other exhortations to increase
    quality, work harder, put forth best efforts, reduce costs

    Incentive- and performance-based pay schemes

- Management viewing their job as one of control

- Arbitrary goals for defect-reduction, as though workers had
  the means to achieve them

- Cultivation of an atmosphere of fear[30]:   *30*

    Reminding workers how expensive they are (costs of
    medical, life insurance, and pensions)

    Ranking and penalizing people on basis of their output

    Punishing people for mistakes

    Laying off workers and threatening layoffs

    Threatening use of outside vendors for processes now done
    in-house

    Selling off units that don't meet the imposed objectives

Contrast Deming's emphasis on the system as cause of most
defects, and on the need to drive out fear and eliminate barriers to pride
of workmanship.   *and, nb, the system is*
                                   an artifact of human
6. Separation of Planning and Doing   *evention*

- Separation was applied to jobs, processes, ideas, innovation,
  and improvement.

- Expertise was assumed to be outside of the process in the
  hands of management and experts in scientific management.

- The system produced the rise of specialized supervisory roles
  (Taylor invented eight himself).

- Workers within the process were assumed to be passive and
  ignorant.   *— and (according to 5 above) malicious*

Taylorism in Action Today (Neo-Taylorism)

- MBO—the manager as reviewer/judge
- Divergence between espoused and actual practice
- Separation of learning from working
- Physical separation of management from the activities of the workplace, leading to decontextuallized views and attitudes
- Reliance on decontextuallized views of the process or system, rather than on the actual practice of those situated within it
- Decontextuallized support functions, such as accounting, purchasing, legal, and personnel, allowed to make policy
- Managers who are removed from both process and product
- Reliance on formal descriptions of work and jobs prepared by "experts" outside of the process
- Adoption of a growing role as a broker for someone else's products in preference to being a producer

content-free management

the virtual corporation.
COTS

Contrast Deming's emphasis on the value to the company of knowledge workers carry around in their heads, and the duty of management to understand the processes they manage and help employees increase their knowledge for improvement of products and processes.

7. Failure To Recognize Systems and Communities in the Organization

we're not in this together

31

- Workers were considered only individually. ("Whenever it was practicable, each man's work was measured by itself."[31])
- No allowance was made for interaction with other workers or with the system.

Taylorism in Action Today (Neo-Taylorism)

- Work viewed individually instead of collaboratively
- Insistence on a unitary, canonical culture, combined with failure to recognize informal communities in the organization
- Employee motivation schemes such as posters, programs, exhortations, and prizes

HR

- Individual power and "perks" offered as motivation, especially for managers
- The rise of the cowboy, or hero, role to deal with crises *Advanz "implemtation team"*
- Policies put people and departments in competition with one another, such as MBO, quotas, ranking (win–lose model)
- "Empowerment" and "self-directed team" programs which only maintain power in management's hands, giving employees a double dose of deception:

  You'll be expected to take responsibility and control of your work." (Translation: "We still control everything. The only thing that's changed is the risk you take.")

  "Your teams will provide their own direction." (Translation: "But we'll be rating each of you, so don't expect cooperation among the members.")

- Ongoing managerial abdication of responsibility for the welfare of employees, as seen in such phenomena as:

  Layoffs

  Various schemes to avoid or limit payment of employee pensions

  Age-discrimination suits

  Employees forced to pay for their own re-training

  People who are below-average in one job released without giving them the chance of reassignment to another job

  Adoption of the "portfolio career" model[32]   *32*

  People whose skills are no longer needed released instead of retrained

  Virtual corps become Virtual corpses   Suppliers used for doing what was formerly done in-house on the pretext of saving money, instead of improvement of internal operations

  Reorganization into new companies in which existing employees lose their seniority

- Suppliers and customers treated as outsiders and regarded with suspicion

Deming is specifically against the above practices, stating instead that:

> *Management's job is to optimize the system,[33] which includes employees, suppliers, and customers, on a win–win basis*
>
> *Management's job is to help their people, to provide security, and to remove the barriers to their taking joy in their work*

[handwritten margin note: 33. But do the Japanese really believe this? Not in any experience I've had or heard of.]

8. View of Workers as Interchangeable, Bionic Machines

- Taylor paved the way for industrial acceptance of John Dewey's theory that man is merely a machine. (Thought, for Dewey, was merely a function of the mechanism called the brain; human beings were "talking organisms.")

[handwritten margin note: but a special machine. AND the machine in charge.]

- Taylor also set the stage for industrial application of the stimulus-response behaviorism (materialistic determinism) of B. F. Skinner (American psychologist, 1904–1990).
- Pay was assumed to be the predominant motivator, especially for lower-level employees. (Taylor's idea was later adopted into a much more elaborate scheme by the behavioral psychologist Abraham Maslow [1908–1970].)

[handwritten margin note: but for maslow, there are many motivators]

- Control of the process was removed from the worker and the traditional foreman.
- Tasks were transferred from skilled to un- or semi-skilled workers.

[handwritten margin note: dummied down]

- No recognition was given to interaction among workers, or to the system in which they work.

[handwritten margin note: neither for social nor productive]

- Combined with a disregard of variation, differences in performance were treated as inherent in the worker rather than caused predominantly by the system in which he or she works.
- If a worker won't do what you want him to, "make him," suggested Taylor

Taylorism in Action Today (Neo-Taylorism)
- Failure to recognize the major effect of the system upon an employee's performance
- Quotas and work standards
- Planned long-term overtime   — *e.g. the UPS strike*
- Management use of the motivational schemes of behavioral psychologists, such as pigeonholing employees according to their economic and organizational status
- Stereotypical organizational responses such as reorganizations which shuffle people around on the assumption that workers, and especially managers, are interchangeable   *no more than secretaries*
- Management giving out rewards and prizes instead of providing intrinsic motivation (joy in work)
- Downsizing to avoid the costs of retaining loyal, effective employees

Here again Deming's condemnation of the following has been outspoken and unequivocal:[34]   *34*

> Forced distribution of grades in school
> The so-called merit system on the job
> Judging people, putting them into slots
> Induced competition among groups
> Incentive pay
> Pay for performance (performance cannot be measured)   *what's the trick here?*
> Numerical goals without the methods to achieve them   *probably op def*
> Explaining variances from goals and targets, leading to   *op def of "perf"*
>   humiliation, fear, defense and competition, playing to
>   win; the crushing of cooperation, joy in work, curiosity
>   and innovation

Each of the above flaws of Taylor's scientific management is serious, and each has serious consequences. The additive effect of all of the flaws together—or of any combination of them—is devastating to an organization whose paradigm is paralyzed in the Taylor mode. Worse, in a kind of *paradigm anesthesia,* most organizations are unaware of the roots of their management philosophy, as scientific management has been declared a dead issue in most business schools and journals.

Philosophical Basis of Taylorism

We have noted how Frederick Taylor walked in the footsteps of earlier men of business. Let's look briefly into some of Taylor's scientific forebears.

The most powerful among these was Sir Isaac Newton (1642–1727), who viewed the universe as a mechanism that followed absolute, determinable laws. Newton's mechanistic orientation reflected the shift from theocentric to secular philosophy as the basis for scientific enquiry, a major feature of the Renaissance. A century later, Pierre Simon Laplace (1749–1827) drew the logical conclusions from Newton's theories and referred to the universe as a gigantic, totally predictable clockwork. According to this view, the future is predictable because it is determined by a combination of an initial state plus the rules or laws of transformation—all of which are assumed accessible to human enquiry.

Twentieth Century Science Shifts Paradigms

Newtonian determinism dominated the science of two centuries, its influence checked only by the emergence of twentieth-century thinkers such as Albert Einstein (relativity theory), Max Planck (quantum mechanics), Werner Heisenberg (uncertainty principle), Henri Poincaré (chaotic processes), C. I. Lewis and Percy Bridgman (epistemology), and Walter Shewhart (statistics).[35] The implications of the activities of these men and their contemporaries were not limited to their own fields. Thanks to the new knowledge they provided, both the scientific method and the philosophy of management were permanently changed, in an inescapable progression. This was a real paradigm shift, signaling an opportunity for those who recognize it, and an assurance that those who don't will be undone by competition that does. Here is the general progression of the shift:

1. *Certainty* about our knowledge of reality was severely weakened. What is left is theory supported by a relative *degree of belief* made possible by operational definitions[36] and experimentation. In addition, some knowledge may be permanently beyond our reach.[37] Deming uses the term "unknowable"; again and again he asks, "How would you know?", challenging the assumptions of people as to what they believe is certain.

36

2. *Causality*—the ability to say conclusively that effect *y* was caused by factor *x*—was severely weakened. Given the present state of a complex system—even assuming it could be sufficiently known—predicting with certainty a future state of the system would be impossible, and equally impossible would be determining a previous state. This is why Deming says that imitation of what seemed to work elsewhere gives no assurance that it will work in your situation.

3. The *nonlinearity of system dynamics* was admitted: systems do not respond in a linear fashion to changes in their inputs, processes, or environments—instead they may exhibit unpredictable, and possibly unpleasant, behavior. Nonlinearity makes even simple systems impossible to predict. Fred Brooks illustrates some effects of nonlinearity in *The Mythical Man-Month*, which draws lessons from his experience managing the early development of IBM's first modern computer operating system, OS/360. Brooks observes, for example, that putting more people on a late project will almost guarantee that it will take even longer.

complexity / chaos theory

but he can explain why?

4. The *nature of knowledge* changed from absolute to probabilistic or statistical,[38] and from fixed to changing; it was broadened to include local and subjective knowledge as

well. In his 1892 book *The Grammar of Science*, Karl Pearson points out that

> *Proof, in the field of perceptions, is the demonstration of overwhelming probability. Logically we should use the word "know" only of conceptions, and reserve the word "believe" for perception.*[39]

This is at least part of the reason why Deming warns that the most important knowledge needed to improve a system is unknown and unknowable.[40] Bringing a process into statistical control may be seen as the key to widening the application of one's knowledge about it via the scientific method.

5. *Variation, randomness, and complexity* were admitted to be inherent to *all* processes, including those of measurement, because processes are all subject to the effects of a constant system of chance causes. Instead of a clockwork, the model of the universe would now have to include stochastic processes—random or chaotic processes which tend as well to produce long stable runs. Because of complexity there can be no "one best way"; optimum processes cannot be designed. This means there is no set of rules and systems management can create that could be guaranteed to improve a situation. Quality gurus who promise certain results from installing their programs have failed to understand this limitation (Deming's profound change 1).

6. *Repeatability of processes* could no longer be assumed because of inherent variation and the local and changing nature of knowledge. It's popular today to define insanity as "continuing to do the same thing but expecting the results to be different," but how could we expect any other result? If the system is unstable, the results will vary greatly even with "identical" inputs and conditions. If it is stable, common-cause variation guarantees randomly-varying results within a

somewhat predictable band. The two cases differ only in
degree.

7. Because *the most important information is not only unknown,
   but unknowable,* in the face of variation and uncertainty
   processes can only be improved, never optimized. The path
   to improvement is via constantly gaining new knowledge
   using the scientific method or the Shewhart PDSA cycle.
   These allow local and subjective knowledge to be widened in
   scope and application through experiment (Deming's
   profound changes 2–4).

These changes tell us *why,* as Konosuke Matsushita has said, the
system of Taylor is dead. And they tell us why making Deming's profound changes is the way to take advantage of this great paradigm shift.

An Unmetaphysical Century

Over the last century, business has changed its foundations from
the rationalism and scientific materialism of the eighteenth century,
which culminated in Taylor's scientific management. But rather than
adopt the system of W. Edwards Deming, business has embraced positivism and Social Darwinism. *Positivism,* associated with Auguste
Comte (1798–1857), stresses the validity of what is observable and dep-
recates speculation about any underlying reality. In "pretending to
know that there is nothing in the universe that could not be investi-
gated and fully clarified by the experimental methods of the natural sci-
ences"[41] positivism takes the arrogance of Laplace a step farther. As one
historian puts it,

*In the broadest sense, it might be said that modern Western
civilization is positivistic, in that metaphysical or religious modes
of thought are not congenial to it. Everyday life is so surrounded
with the technological and the scientific, so extensively "rationalized," so conditioned to mechanical models and explana-
tions, that conscious mental life runs naturally and normally in*

grooves than can be called "positivist," that is, scientific, rational, nonmetaphysical, averse to mysticism or any truths not immediately verifiable by experiment or demonstration.[42]

Not surprisingly, the positivistic frame of mind, with its obvious corollary of gaining power through a showing of evidence, has led to the current trend of "proving" claims by citing evidence which is outright spurious, soon contradicted, or ignores most of what we describe here as the scientific method. The controversies rage over health claims, environmental claims, legal claims, claimed causality between anything and anything else. In place of the methods of science, a mere emotional appeal by those who imagine themselves "victims" is often enough to bias the powerful media in their favor.

Managers in both business and government often use such causes to enhance their power and justify their suboptimizing decisions. Some of the more egregious pretexts may be the assumptions—without proof—that employees are lazy and work only for material rewards; and that they therefore must be told what to do, scientifically manipulated, and got rid of when business turns down.

Social Darwinism, a term attributed to Herbert Spencer (1820–1903), is an adaptation of Darwin's "survival of the fittest" theory to human societies. Extending, as did Marx and Darwin, the Hegelian dialectic of thesis/antithesis/synthesis, Social Darwinism supplies many of the metaphorical perspectives of hero and adventurer with which business leaders often identify themselves and their organizations. To this view, competition is the arena in which the theses and antitheses of the participants are resolved into syntheses in which the fittest prevail, contrasting sharply with the economic view of competition as a continual process of discovering consumer needs in order to serve them.

In the preceding list of flaws of Taylorism and neo-Taylorism, there were many instances of both positivism and Social Darwinism, as shown in the matter-of-factness with which businesses are disposing of what they formerly considered assets, such as people and relationships, and filling their places with the search for market share and high stock prices. "Unless you leave we won't *have* a business," is spurious reasoning which is often a preamble to layoffs and buy-outs; this clearly reveals the shift in the world-view of business that has occurred over the last century, from accountability to people and relationships to accountability to no one except some vague, inevitable, future Darwinian corporate synthesis.[43]

In Case You Missed It

Organizations enmired in neo-Taylorism are the victims of paradigm paralysis, clinging to a school of management that is not only outmoded for present times but has been grotesquely transmogrified into a self-destructive philosophy which is ravaging people, companies, and the industrial strength of the West. Those organizations that don't admit this fact and transform their philosophy will be eclipsed by those that do.

NOTES

1 The Fourteen Points are listed in Appendix B.

2 From J. Anderson, K. Dooley and S. Misterek, "The Role of Profound Knowledge in the Continual Improvement of Quality," *Human Systems Management* 10 (1991), pp. 243–259.

3 When we ask people who have attended Deming's seminar what he was trying to teach them, they rarely give any of the profound changes. When we explain to them what these changes are, they remember spending lots of time on them—and they could appreciate the importance of them—but in class it had often gone in one ear and out the other.

4 Many people from industry and the universities contributed to the training for supporting the war effort. Much of the training effort was limited to imparting and improving techniques of inspection for obtaining quality goods. For a discussion of Training Within Industry, one of the more lasting war-training efforts, see Alan G. Robinson and Dean M. Schroeder, "Training, Continuous Improvement, and Human Relations: The U.S. TWI Programs and the Japanese Management Style," *California Management Review* (Winter 1993), pp. 35–57.

5 Shewhart and Deming call this—reacting to noise as if it were a signal—mistake number one. Mistake number two is treating signals as if they were noise—ignoring a signal. Because of the limitations of what can be known, which we will be discussing shortly, management can never reach a state where neither of these mistakes is ever made. The best they can hope for, Shewhart says, is to make mistake number one once in a while, and to make mistake number two once in a while. It is futile to hope for this achievement without understanding the theory of variation and the use of control charts.

6 For clarity, here are the definitions for system and process that we are assuming in this book. A *system* is a regularly-interacting or interdependent group of parts forming a unified whole and serving a known aim so as to benefit all of its elements. A *process* is a series of actions of operations conducted towards an end.

7 PDSA (Plan-Do-Study-Act) is a business application of the scientific method. It can be traced to the work and writing of Walter Shewhart, Deming's greatest mentor, in the 1930s. By iterating the PDSA cycle an organization can continuously gain the new knowledge needed for improvement.

8 Extension transferences will be dealt with in section 4.2, "Phases of Learning."

9 Kosaku Yoshida, a native-born Japanese who now teaches in the U.S., explains these characteristics and suggests why Westerners must adopt them to some degree if they expect to undergo Deming's transformation, in "Deming Management Philosophy: Does It Work in the US as Well as in Japan?" *Columbia Journal of World Business* (Fall 1989), pp. 10–17.

10 We are indebted to Mr. Barry G. Carroll of IBM for these insights.

11 In their paper "Continuity and Change in Japanese Management," Tomasz Mroczkowski and Masao Hanaoka cite trends toward adoption of such Western practices as performance-based evaluation and rewards for individuals, and the decline of the seniority system and lifetime employment, noting that Japanese executives increasingly are finding no time for traditional consensus-building. *California Management Review* (Winter 1989), pp. 39–53.

12 A few years ago Perry Gluckman and Marian Hirsch wrote a paper entitled "Is Taylorism Slowing Competitiveness and Productivity Improvement?" Of twenty journals to which Perry sent the article, sixteen rejected it on the basis that Taylorism was dead and that the article was beating a dead horse; none of the other four replied. Perry presented the paper at the University of Chicago's Graduate School of Business, and was assured by the assembled professors that Taylor's influence was indeed long dead. At this point a graduate student dissented, saying that everything Perry had attributed to

Taylor was being taught in the books assigned for the courses she was currently taking, and offered page references!

13 From Drucker's essay, "Frederick Taylor." Quoted in David A. Whitsett and Lyle Yorks, *From Management Theory to Business Sense: The Myths and Realities of People at Work*, AMACOM, 1983.

14 As Walter de Gruyter notes in *The Japanese Industrial System* (Charles J. McMillan, 1985), scientific management was under study in Japan as early as 1908, and major Japanese industries were adopting it by 1915. The Taylor system probably prevailed in Japan through World War II.

15 From Robert Owen, *A New View of Society*, E. Bliss & E. White, 1825.

16 Published by Carey & Lea, 1832.

17 Taylor later said he was introduced to time studies by Dr. Wentworth, one of his teachers at Exeter.

18 Taylor discusses soldiering practices in the first chapter of his work *The Principles of Scientific Management* (Harper & Bros., 1911). Soldiering—where it actually did happen—was, in the opinion of the scholars of this period, not a reflection of laziness but a reaction to the widespread management practice of reducing hourly rates whenever the workers became more productive in order to shave costs. See, for example, Aitken, *Scientific Management in Action* (note 20).

19 *The Principles of Scientific Management*, p. 52.

20 Per Hugh Aitken, author of *Scientific Management in Action: Taylorism at Watertown Arsenal, 1908–1915* (Princeton University Press, 1985 [1960]), "The apparent accuracy and objectivity of stop-watch time study was therefore to a large extent an illusion. When a task time was set for a certain job, one part of the total had been set by reading measurements from a precise instrument—the stop watch. The other part had been set by a whole series of conventional decisions, in which the values and preconceptions of the individual doing the timing were foremost" (p. 26).

21 See Taylor's paper "A Piece-Rate System, Being a Step Toward Partial Solution of the Labor Problem," *Transactions, A.S.M.E.*, 16, 1895.

22 Taylor's *Principles* was published in Japan under the name *The Secret of Lost Motion* and sold two million copies, building upon the momentum begun there a few years earlier.

23 In April 1918, Lenin declared, "We must introduce the study and teaching of the new Taylor System and its systematic trial and adaptation." See Edward E. Hunt, editor, *Scientific Management Since Taylor*, McGraw-Hill, 1991, p. xi.

24 From *Science in Management for the One Best Way To Do Work*, a paper Gilbreth presented in Milan in 1923.

25 From *Freedom and Coordination*, Management Publications Trust, 1949.

26 John H. Williams, "A Technique for the Chief Executive," *Bulletin of the Taylor Society* (April 1922), 7, No. 2.

27 De-skilling and its implications are discussed by Helga Drummond in *The Quality Movement: What Total Quality Management Is Really All About!* (London: Kogan/Page; U.S.: Nichols, 1992, pp. 139–141). A major problem with de-skilling is achieving any persuasive degree of belief that, in doing a task that is well beneath his or her skills, the more skilled worker adds nothing significant beyond what a less skilled one would. The tacit, subjective knowledge—the "knack"—in the mind and hands of the worker is just as much a part of the process as what has been objectified and reduced to explicit procedure. Also see Professor Ikujiro Nonaka's article "The Knowledge-Creating Company," *Harvard Business Review* (November-December 1991). pp. 96–104.

28 See Hugh Aitken, *Scientific Management in Action* (note 20 above), for a discussion of the idealized or notional characteristics of time measurements as conducted by Taylor.

29 Suboptimization is in no way limited to business: elected representatives to the federal government typically think their chief purpose in being in Washington (after being re-elected, of course) is to secure or hold onto federal spending in their home states or districts—providing us with two examples of suboptimizing behavior.

30 In their book *Driving Fear Out of the Workplace*, (Jossey-Bass, 1991), Kathleen Ryan and Daniel Oestreich provide a useful list of the ways in which fear is generated, and manifests itself, in an organization.

31 From *Shop Management*, a paper Taylor presented at the June 1903 meeting of the American Society of Mechanical Engineers.

32 The portfolio career model is a new job relationship that allows the employer to treat the employee as though he or she were an outside supplier instead of a member of a community. Such relationships are characterized by lack of benefits, pension, or employee development. The innocent-sounding name tends to legitimize this highly suboptimal and damaging practice.

33 An optimum system would be one which has achieved its aim. Because of variation and the attendant imperfection of information, the "optimum" state serves as a goal that is approached incrementally in an indefinite series of steps involving new levels of order and knowledge (e.g., PDSA). This latter journey is what Deming means when he says management's job is to optimize the system.

34 Cited in his George Washington University satellite broadcast of April 1992.

35 Note to scientists: of course most of Newton's laws, such as gravitation and

motion, remain intact today—for most purposes. What modern physics did to Newton was not to discard him but to show that his ideas did not have the universal applicability previously believed, as well as to discover new laws which successfully undermined the assumption of mechanistic predictability with which he and his followers viewed the universe.

36 The physicist Percy Bridgman, a leading advocate of the school of operationalism, insisted that there were no true values; there were only the results of carrying out a procedure of measurement. We see what Deming learned from this in his insistence on operational definitions in business.

37 In his famous paper, delivered in 1927, Heisenberg said, "In the strict formulation of the causal law [of Laplace]—if we know the present, we can calculate the future—it is not the conclusion that is wrong but the premise," meaning our knowledge of the present must be "in general only of a statistical type." From David C. Cassidy, *Uncertainty—The Life and Science of Werner Heisenberg*, W. H. Freeman, 1991, pp. 228–229.

38 "Today the mathematical physicist seems more and more inclined to the opinion that each of the so-called laws of nature is essentially statistical, and that all our equations and theories can do, is to provide us with a series of orbits of varying probabilities." From the July 1927 issue of *Engineering*, quoted by Walter Shewhart in *The Economic Control of Quality of Manufactured Product* (American Society for Quality Control, 1980). See also the above quotation from Heisenberg, in note 37.

39 London: J. M. Dent & Sons, p. 130.

40 Does this mean the external world is essentially unknowable—that, contrary to Einstein's belief, God really does play dice? We prefer the following: "When God created the universe, He also ordained mathematical relationships between all things then existing and all things that ever would exist. In conducting an experiment the scientist is trying to determine the values that God ordained for $f(x)$, $f(y)$, etc. Almost certainly, the right answer will not be obtained. All the experimenter can hope for is that the answer obtained from the experiment will be a good estimate of the value in God's equation." From the introduction to William J. Diamond, *Practical Experiment Designs for Engineers and Scientists*, van Nostrand Reinhold, 1981.

41 From Ludwig von Mises, *The Ultimate Foundation of Economic Science*, Sheed Andrews and McMeel, 1962, p. 53.

42 Roland N. Stromberg, *An Intellectual History of Modern Europe*, 2nd ed., Prentice-Hall, 1975, p. 302.

43 John J. Clancy treats this subject with perception and excellent research in *The Invisible Powers: The Language of Business*, Lexington Books, 1989.

Understanding Deming's Profound Changes

What are Deming's profound changes? Why they are important, and what benefits do we get by adopting them?

Deming's profound changes, as discussed in detail in the following chapters, are related to increasing a company's affordable growth rate, increasing quality and productivity through division of labor, and learning new knowledge faster. Both the obvious benefits, as well as some not so obvious, are discussed. First we list the profound changes and explain each, relating its importance to any organization interested in survival—and beyond that, in growth. We then begin the enumeration of the benefits of the profound changes, a discussion which carries on through the remainder of the book.

3.1 What Are Deming's Profound Changes?

Managers always believe they are doing the right thing regardless of the style of management they practice. They think all they have to do is to work harder at it. This is not true. It is vital for them to recognize the need to reduce variation and its consequences, as well as the need to offer customers the grade[1] and value that they require in products and services, and the need to treat employees and suppliers as part of their system.

Here are the changes that Deming taught the Japanese that Western management still need to learn:

1. **Every system has variation; hence, the information needed to create optimum systems is unknown and unknowable.**

Notes for this chapter begin on page 58.

2. Using the scientific method we learn what's unknown but knowable faster.

3. By observing the operation of the system, built-in flaws can be detected and isolated.

4. Complexity can be reduced and entropy lowered by removing the built-in flaws.

System Variation

Because every system has variation, the first profound change says that all systems will have built-in flaws. The shift from Newtonian mechanics to quantum physics, and our modern understanding of special relativity, have changed more than our approach to science: They have changed the ways we must do business.

Sir Isaac Newton is known for many laws and discoveries, as well as his presupposition that the universe could be described by a series of equations which were known, or were accessible through analysis. However, twentieth century physicists have discovered that the world is not so easily defined, with the greatest work having been done in quantum physics. These discoveries challenge the validity of applying Newton's theories to all physical phenomena we observe today. Newton's theories were adequate for his time, but as time and experience provide us with more knowledge, we find previous models do not explain all of what we observe. The old models still have limited function, but new ones must be developed to explain new discoveries. We see a similar emergence of a new set of models as Deming's theories eclipse Taylor's.

One of the most important of Deming's changes is the recognition—*virtually unique with Deming among all the management gurus*—that the information necessary to optimize any system is not only unknown at this time, but inherently unknowable to us. This also means that, because any process may be affected by inputs that we either have no information about or aren't aware of at all, expecting that process to perform identically in another place or time would be foolish.

This recognition of the unknowability of system-optimizing information, with its attendant loss of predictability, is one of Deming's greatest contributions to business, counterbalanced by his admonition to do what we *can* do—keep learning and improving our systems with the new knowledge. In turn, Deming gives much credit to Shewhart whose pioneering work on the difference between noise and signals, and its already discussed implications for management action, presaged all four of these profound changes. Taylor recognized variation, but attributed it to the workers and dreamed of eliminating it by means of his time-and-motion studies, work standards, and differential pay schemes.

Aside from the occasional, infrequent opportunity for true breakthroughs, our day-to-day choice in the face of this limitation is to pursue *incremental improvement* through the learning process known as the scientific method. Taylor's writings reveal that he was unaware of this. Although he believed in improvement and was no champion of complexity, he does not appear to have understood the concept of variation, the use of the scientific method, or the need to continually reduce complexity.

The Scientific Method

The scientific method is used to discover new knowledge. Walter Shewhart introduced the business world to using the scientific method on its own operations in 1939 when he showed that the three steps of specification, production, and inspection, which were usually thought of as following one another in a straight line, should instead go in a circle or spiral:

> It may be helpful to think of the three steps . . . as steps in the scientific method. In this sense [they] correspond respectively to making a hypothesis, carrying out an experiment, and testing the hypothesis. The three steps constitute a dynamic scientific process of acquiring knowledge. From this viewpoint, it

*might be better to show them as forming a sort of spiral grad-
ually approaching a circular path which would represent the
idealized case where no evidence is found in Step III to indicate a need for changing the specification (or scientific
hypothesis) no matter how many times we repeat the three
steps. Mass production viewed in this way constitutes a continuing and self-corrective way for making the most efficient use
of raw and fabricated materials.*[2]

Note: The scientific method was almost unknown in
Japanese business before Deming taught it there in 1950.

As the statistician George Box observed, two elements are necessary to discover something new: a critical event and a perceptive
observer.[3] Critical events are inconsistencies or contradictions between
what is predicted and what is observed. Whether planned or
unplanned, these events contain significant information for the perceptive observer.[4] These are the clues to nonlinear response and synergistic
effects among the various components of the system—effects that offer
unusually great leverage if used to improve the system.

The scientific method is a path toward what is unknown but
knowable. Because it is much more effective at discovering new knowledge than are guesswork and imitation, using the scientific method in
business can speed up learning new technical knowledge, the foundation for new product and process technology; it can also provide the
basis for continued reduction of variation even in stable systems by
using the leverage of demonstrated system nonlinearities.

But, in order to be perceptive, an observer must first have a theory. Without a theory the observer has only his or her undifferentiated
experience from which to learn something new. As Deming says flatly,
"Without theory, experience teaches nothing. In fact, experience can
not even be recorded unless there is some theory, however crude, that
leads to a hypothesis and a system by which to catalog observations.
Sometimes a hunch, right or wrong, is sufficient theory to lead to useful
observations."[5] Without theory the observer simply cannot observe—
he has no way to select certain phenomena in preference to others

among the nearly infinite variety available to his perception.[6] As Einstein said, "It is the theory that decides what we can observe." So, for example, some organizations keep track of defects as a key measure of customer satisfaction. This implies a theory that defects are of key importance to customers, and correlate well with their continuing to buy and use the product or service. But some customers may value timeliness of delivery more highly than defect-free product; unless the producer changes his theory he has no way to observe in the same manner as those customers do.

Deming has been saying for at least ten years that management must have deep, profound knowledge of their subject matter—the processes that create their organizations' products and services. So, to be good theorists, management must intimately know their system and its aim before they can generate useful theories for increasing knowledge about them.

One may theorize—using process-based intuition, for example—that the yield from a process is the same at all times of day (expected response). By observing the process and by using statistical analysis, he may find that yields differ considerably between first and second shifts (actual response). He is now in the position where he may develop a new theory that accounts for these differences—for example, by the discrepancies in training in a soldering operation. Then he may, by removing the cause of the discrepancy, create the conditions under which the original assumption would be true.

System Observation

By observing statistically a system while it is in operation—as opposed to Taylor's approach of having a team of experts work on the system off-line from its operation—we can, by using the scientific method, detect and isolate the built-in flaws. By *observing statistically* we mean using a time-ordered series of measurements from the system, which are analyzed using the criteria for the state of statistical control.[7]

Once isolated, the flaws can be removed, which will reduce the variability and lower the entropy of the system. Once a system is in statistical control, designed experiments can reveal the leverage points for removing common causes of variation in order to maximize the system's robustness and performance.

Surprisingly, one of the more significant early benchmarks in quality-improvement efforts, especially in the service sector,[8] is acknowledging that one's work is a process involving repetitive steps. Often we hear, "We don't have a process" to do some task—yet clearly the job gets done again and again. If the work is being done, it is by means of a process; this is definitional.

Complexity

Reducing complexity is the key to improving productivity, increasing profitability, and having a higher grade of product or service. By *complexity* we mean higher degrees of randomness, disorder, and unpredictability. Complexity in a process, for example—whether that complexity is inherent or is unnecessary and arbitrary—leads to greater variation in its output. Complexity is a *cause* of problems—but few causes are simple, first-order causes: They are usually themselves the products of complicated interactions. Thus, complexity is both a cause and an effect of variation.

The terms complexity, entropy, and variation are generally nearly synonymous in this book. We will illustrate why, beginning with the concept of entropy:

> The quantity of disorder is measured in terms of entropy. One way of defining entropy is in terms of the number of states, or degrees of freedom, that are possible in a system in a given situation. The disorder arises because we do not know which state the system is in. Disorder is then essentially the same thing as ignorance.[9]

The greater the number of states a system could have, the greater its complexity. Variation in a process means that its outputs can have a number of states, rather than just the target state that its customer wants. Table 2 shows some important types of variation along with some of their characteristics.

Order and predictability are not natural states; they must be imposed by the application of energy from outside the system. Left to themselves, isolated systems will "run down"—disorder will increase spontaneously toward an equilibrium of maximum entropy. This is the Second Law of Thermodynamics. Sir Arthur Eddington noted in 1927 that this law "holds the supreme position among the laws of Nature . . . If your theory is found to be against the second law of thermodynamics, I can give you no hope; there is nothing for it but to collapse in deepest humiliation."

Type of Variation	Comments
Deterministic (systematic)	Traceable to specific factors, such as change of shift or replacement of tool. Often it's the result of policy.
Common-Cause	Composed of random and systematic variation. Often it's persistent in time and diffuse in effect. Each time a stable process is executed, a variable result is produced! The difference between subsequent data points in a process in the state of statistical control can be thought of as *noise*.
Special-Cause	Traceable to specific factors outside of the process. Typically it's sporadic in time, local in effect. A special cause is a *signal* that something has changed, that a rare event has occurred.
Chaotic	Complex behavior may be exhibited by a phenomenon explainable by a simple, but perhaps unknown, mathematical model. Very sensitive to initial conditions.

Table 2. Types of Variation

3.2 Shewhart's PDSA Cycle

An exciting prospect for increasing knowledge through the scientific method is the embodiment of the new knowledge in technology that can actually reverse entropy:

Take an electric clock. It converts electrical energy into mechanical energy via an electric motor. . . . A traditional analysis would consider the work output to consist of (force x distance) moving the hands of the clock. A side product would be heat, and the entropy of the universe would have been increased. Yet the clock also provides information. It may be used as a timer to turn on a video recorder, turn off a microwave oven, or operate delicate machinery in a factory . . . . The output of a station clock may therefore decrease the entropy in the universe by organizing a transportation network . . . . Thus machinery not only contains information, but part of the work performed by machines involves the creation of new information.[10]

This surely helps to account for Deming's insistence that there is no substitute for knowledge—profound knowledge that is constantly being expanded by learning and discovery, and encoded in entropy-reducing technology.

Figure 1 illustrates the relative movement from entropy and igno-

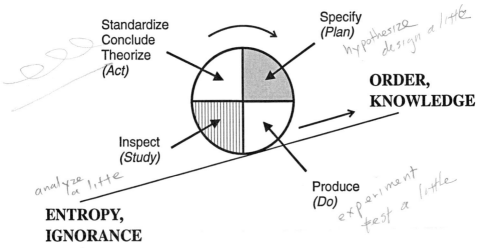

Figure 1. The Shewhart, or PDSA, Cycle.

rance toward order and knowledge by means of continuously repeating Shewhart's cycle of gaining new knowledge and standardizing it.

Shewhart called his cycle "a dynamic scientific process of acquiring knowledge."[11] He applied it to consumer research and product improvement, as well as to process improvement. We begin with a theory or hypothesis, including a procedure for testing it (Shewhart's product specification). We then proceed to carry out the experiment (Shewhart's production step), whereupon we study the result, including data (Shewhart's inspection step, including customer feedback). Our study leads us to various kinds of action: to making new conclusions about our theory, or to standardize our process or product at the new level, or to planning another experiment (process or product redesign).

The Taguchi Loss Function[12] tells us that variation of output from the target value of the process causes economic loss to the producer, which the producer typically passes on to the consumer and thus to society in general. Furthermore, this loss is nonlinear and begins with the slightest variation. Deming amplifies this by emphasizing that centering the process on the target value is at least as important as reducing variation about the process average. The way toward this kind of real improvement is by increasing order and knowledge. Because all observed phenomena are subject to constant systems of chance causes, our knowledge about them is never perfect. The uphill path to new knowledge and higher degrees of order is ascended by iteration after iteration of the Shewhart cycle.

3.3 Why the Profound Changes Are Important

Decades ago, David Packard, co-founder of Hewlett-Packard, presented what he considered some of the fundamental truths about expanding a business. He explained these as elements of an equation. Packard's equation is as fundamental to running a business as Ohm's law is to electrical engineering:

The percentage increase in sales which you can finance

> *each year is equal to your percentage of profit after taxes times your capital turnover. Capital turnover is defined as the dollars in sales you can produce per year for each dollar of capital you have invested in your business. . . . Your capital includes working capital (that is the money used to buy inventory, to finance your accounts receivable, to provide some working cash, etc.) and fixed capital would be the amount of money you have spent to buy facilities, tools and equipment.[13]*

Packard's equation thus provides a working relationship between management policy and growth of the business. He puts the concepts of return on sales and asset turnover into a form that even non-accountants are able to appreciate.

Some non-affordable alternatives to growth which many companies have chosen to their detriment include going into debt and issuing stock. As Henry Ford observed two generations ago, the soundest way to increase growth is to use the productive capacity of your operations, which keeps your company free of control by creditors or the stock market:

> *[On borrowing.] The place to finance a manufacturing business is the shop, and not the bank. . . . The money has to come out of the shop, not out of the bank, and I have found that the shop will answer every possible requirement.*

> *Money is only a tool of business. It is just part of the machinery. You might as well borrow 100,000 lathes as $100,000 if the trouble is inside your business. More lathes will not cure it; neither will more money. Only heavier doses of brains and thought and wise courage can cure. A business that misuses what it has will continue to misuse what it can get. The point is—cure the misuse. When that is done, the business will begin to make its own money, just as a repaired human body begins to make sufficient pure blood.*

> *[On stockholders.] The stockholders, to my way of thinking, ought to be only those who are active in the business and who will regard the company as an instrument of service rather than as a machine for making money. . . . Hence we have no place for the non-working stockholders. . . . If it at*

*any time became a question between lowering wages or
abolishing dividends, I would abolish dividends.[14]*   14

By reducing complexity most companies could produce more than
twice as much product or service in less than half the time. This remark
may seem startling at first, but it is readily understandable. Most companies have bottlenecks in their production systems that operate at
50% of capacity; once these are identified, the entire flow can be paced
by the bottleneck, leading to reduced inventories, and to freeing up
wasted capacity for holding and accounting for work in process.
Packard's equation says that, because you are now using fewer assets to
reproduce the same amount of goods, you have now increased your
affordable growth rate.[15]   15

Increasing the division and concurrency (simultaneity) of work—
and gaining knowledge faster to improve the process and make better
predictions—are two important driving forces of Western culture that
have repeatedly created new, improved ways of producing goods and
services beyond the methods used previously. Division of labor leads to
(1) quicker advancements due to focus and specialization, and (2) the
opportunity for concurrency of work. Changes of this significance are
called *paradigm shifts*.

Moving from Taylor's school of management to Deming's school
of management requires making at least two paradigm shifts or extensions: first, from Newtonian to quantum physics, and second, from scientific management to scientific method. Failure to make these shifts
will deprive the West of two of the most important advances in management philosophy in the last several centuries, and will keep it powerless to deal with whoever adopts the more advanced thinking. We *when*
could lose not only industrial power but also our national sovereignty.

Implied by Konosuke Matsushita was that we are committing
self-destruction by keeping our heads Taylorized. Current management
under neo-Taylorism is, unhappily, a stable system; it must undergo
transformation before the heavy losses it has caused can be reversed. Is
it any wonder that we ask when the sleeping giant will awaken?

Benefits of Deming's Profound Changes

The advantage of understanding and using Deming's profound changes is that improvement of the grade of product or service for the long term is easier, with minimum short-term reduction of profit margin.

Two hundred years ago, in his book *The Wealth of Nations* (1776), Adam Smith (1723–1790) observed that division of labor increases efficiency because: (1) workers can improve their dexterity; (2) less time is wasted in switching tasks; and (3) proper technology can be implemented more effectively within each operation. This is just as true today. Concurrency exploits the division of labor to expand capacity by doing multiple things at the same time.

The major point is that we can shorten the time to produce, supply, or create a product or service. The result is that we can increase our capital turnover, and thus our affordable growth rate.

We live in a world of time-based competition. Usually the Japanese are finished before we can get started. The benefits of doing more with less, having easier to solve problems, and getting at opportunities easier has clear advantages if you want your company to survive. Therefore, using the new paradigm will reward those who understand and lead in the use of the tools that exploit Deming's profound changes. The faster we turn the PDSA wheel, the faster we discover new knowledge for expanding markets and improving systems.

NOTES

1 *Grade* refers to the degree to which additional characteristics are provided beyond the required basics for a product or service. Higher grade usually costs more. A few decades ago Sony used to sell portable transistorized radios; today Sony still offers portable radio reception, but has increased the grade of product by adding features such as headphones, stereo, tape and disk players, miniaturized components, etc. Increasing the grade of product is an effective, system-optimizing competitive strategy which expands markets.

2 *Statistical Method from the Viewpoint of Quality Control*, Dover Press, 1991 [1939], p. 45.

3 George E. P. Box and Søren Bisgaard, "The Scientific Context of Quality Improvement: A Look at the Use of Scientific Method in Quality Improvement," *Quality Progress* (June 1987).

4 Our friend Dr. A. V. Viswanathan of Boeing Aircraft Company notes that contradictions tell you "what you know that ain't so."

5 *Out of the Crisis*, p. 317.

6 The scientific method is not limited to scientists; even accountants use it. When CPAs write an audit report (e.g., the kind that appear in annual statements of corporations) they state that "[our] examination was conducted in accordance with generally-accepted accounting and auditing principles . . . ." This declares the theory that guided their perceptions as observers, in this case the theory that is published by the American Institute of Certified Public Accountants. Without this, or some other, theory, auditors would be unable to audit.

7 Statistical control is taught in several good books, such as those by Shewhart and Deming listed in this book's Bibliography. Also see Donald J. Wheeler, *Understanding Statistical Process Control* (2nd ed.), SPC Press, 1992.

8 Deming often emphasizes that the "service sector" comprises over 80% of America's work force.

9 Stanley W. Angrist and Loren G. Hepler, *Order and Chaos*, Pelican Books, 1973.

10 Tom Stonier, *Information and the Internal Structure of the Universe: An Exploration into Information Physics*, Springer Verlag, 1990.

11 See pp. 45 and 149 of his *Statistical Method from the Viewpoint of Quality Control*.

12 See Genichi Taguchi, *Introduction to Quality: Designing Quality into Products and Processes* (Asian Productivity Organization, 1986) for an explanation of the loss function.

13 From *Growth From Performance*, an address presented by Packard at the Seventh Region Conference of the Institute of Radio Engineers, 24 April 1957.

14 Henry Ford, *My Life and Work*, Garden City, 1922.

15 Further contributions to this theory are given in Eliyahu Goldratt, *The Goal: A Process of Ongoing Improvement*, North River Press, 1984.

Gaining New Knowledge

4.1 Extension Chains and Paradigm Shifts

Organizations can make solid progress by developing and adding to extension chains or by paradigm shifts. As Joel Barker[1] observes, after a paradigm shift occurs everything is reset to zero for everyone, for all employees and all companies within an industry. Deming often cites the example of automotive fuel injection taking over, in just a few years, the job formerly done by carburetors, leaving some carburetor manufacturers wondering what happened.

Paradigm shifts are especially troublesome for companies whose system either has no understood aim, or whose aim is simply to provide a return on investment to shareholders with little reference to the needs of the other elements of their systems, such as employees, suppliers, and customers. Those companies whose management have organized according to Deming's flow diagram of production viewed as a system (discussed later in Chapter 8, Axiom 3, Using Systems) should be in a good position to take advantage of—and even create—the paradigm shifts and extensions that cause the great lunges of progress in industry.

Extension chains and paradigm shifts are quite complementary concepts. The term *extension chain* is taken from the work of Edward T. Hall, a cultural anthropologist. Hall describes an extension as using the environment as a tool, a "whole series of new and often unforeseen environmental transactions that require further adjustments." He notes further that:

Notes for this chapter begin on page 80.

*One purpose of an extension is to enhance a particular func-
tion: the knife does a much better job of cutting than the
teeth . . . language and mathematics enhance certain
aspects of thinking . . . the telescope and the microscope
extend the visual memory system. Extensions often permit
man to solve problems in satisfactory ways . . . but the extension does something else: it permits man to examine and
perfect what is inside his head . . . to look at it, study it,
change it, perfect it.[2]*

Chains of extensions therefore connote the way we make progress,
including its unfinished and sometimes retrograde nature. Because Hall
has made the definition general, almost anything could be an extension:
a tool, a technique, or a philosophy. A paradigm shift can either cause,
or be caused by, an extension.

By failing to make the necessary paradigm shifts an organization
can fall hopelessly behind if it lags on grade of product or service, or
knowledge of new technologies.

The noticeable shortcoming in most of the Fortune 500 companies
is how slow they are to produce new products and knowledge. In the
decade it took GM to roll out the Saturn automobile, Honda—a much
smaller company—introduced four totally new passenger cars,
redesigned every existing car in its line twice, and in addition introduced a high-performance luxury sports car to challenge the Ferrari.
The organizations who best master the art of gaining new knowledge
and can do it faster will be the new Fortune 500 companies.

4.2 Phases of Learning

In business, and indeed in all aspects of life, a person exhibits to
various degrees the trait of *learning*. All too little is commonly known
about how we learn; most people not only approach learning intuitively, but also rarely take any explicit steps to increase the rate at
which they learn. This is very unfortunate in both business and life in
general because the rate at which we gain new knowledge governs our

ability to increase our security and standards of living. We must learn to learn *faster.*

Perhaps the highest rate of learning is achieved in the first few years of life. Children when made aware of their inability to deal with their own circumstances try again and again to master them until they have some effect on them or grow bored. Yet this very method of attacking a problem literally and intuitively causes grief in later life, because people typically do not adopt a *principle of learning* until they are quite mature, and many never do.

In this section we present a system of *phases of learning* which we feel is implied in Deming's philosophy—Deming himself has said that the aim of his philosophy is to bring about "a new kind of world," not just an improvement in quality. Though neither final nor authoritative, these phases of learning are a deliberately developed set of steps to provide essential help in articulating the depth to which one must go in gaining what Deming refers to as profound knowledge. In general, think of the system as a continuum which we have pragmatically divided into bounded component parts for purposes of description. Each phase of learning is given a number to indicate its position in ordinary experience, but there is no guarantee that any given person will achieve all the phases.

The boundaries between phases are continuous rather than discreet, and a person may be in more than one phase at a time. Even reversion to an earlier phase is possible. The first two phases are set apart from the others because they comprise *non-learning* or *pre-learning* phases.

We feel growth through the last four phases (the *learning* phases) is vital to those who are making an effort to improve upon the systems they are a part of. The thread that runs throughout all these phases is the gaining of *new knowledge.* This progression marks the person who is truly a student of Deming and is trying to adopt his changes. This growth may take a lifetime, but it will be worth it!

Table 3 lists the phases in order, and a discussion follows of each phase.

Pre-Learning or Non-Learning Phases

1. *Ignoring Problems.* Turning our backs on a problem in the hopes that it will go away, or does not really exist, usually means that it will be even worse when we finally turn around.
2. *Manipulating the Symbols of Solution.* Superficial or cynical approaches to "doing something" about a problem deny us the opportunity to gain the knowledge required to solve it.

Learning Phases

3. *Solving Problems.* Direct, narrowly focused problem solving approaches may resolve the immediate problem but often ignore systematic causes.
4. *Defining Problems.* The scientific method provides the basis for gaining new knowledge, plus greater awareness of what learning is.
5. *Questioning Problems.* Exploiting the knowledge available from contradictions and inconsistencies between our theories and the real world greatly enhances our ability to improve the system that lies beyond our immediate concern.
6. *Adopting a Principle of Learning.* Recognizing the need for continuous learning and improvement of the system leads toward knowledge which enables us not only to make the system better, but to continually move its boundaries outward.

Table 3. Phases of Learning

1. Ignoring Problems

Some problems are easy to ignore. If no harm appears to come from ignoring a problem, we tend to continue as before. This can be done unknowingly, as in the case of a disease whose symptoms are not recognized; or willingly, as we do so often with conditions such as obesity, smoking, or the problems of business that we are discussing in this book. Why would anyone ignore a recognized problem? Here are some of the reasons.

- You think the problem is smaller than it is, and therefore that it requires no attention.

- You follow a local paradigm that considers the problem insignificant or incapable of solution. This mapping of the symptoms into a frame that doesn't have to be dealt with may be based upon superstition or other lore that simply isn't true, such as Taylor's assumptions that workers are lazy and respond only to monetary rewards.
- You feel you already know how to handle it, and that therefore it isn't really a problem.
- You believe you are in fact handling the problem, but you are deceiving yourself (as Chris Argyris points out, one's espoused theories of action may differ widely from the theories in use.)[3]
- You hope it will go away on its own.
- You hope someone else will take care of it.
- Repeated attempts to attack it have failed, so you redefine it as a non-problem.
- The problem is perceived to be outside your scope of influence and is therefore someone else's problem.

Sometimes our reasoning for ignoring a problem is correct, and someone else does handle it, or it goes away on its own. Other problems only get worse, especially when we underestimate their size or when we continue our youthful habit of relying on a learned repertoire of solutions. For example, when the Volkswagen first came to the U.S. General Motors, Ford, and Chrysler dismissed any possible threat it might bring to their markets because it was so ugly and small, and only a few oddballs were buying it. By about 1960 the popularity of the Beetle was recognized, but Detroit "knew" what Americans *really* wanted was simply a styling change, something it was already very good at. The same scenario was played again fifteen years later, this time with the "different" cars from Japan. Only in the last few years have the American auto makers realized the *quality* of their vehicles was part of the equation and that merely imitating the Japanese or the

Germans is not the answer. If this lesson has indeed been learned, it took an incredible thirty-five years—a period of great cumulative losses.

Ignoring a problem is a frequent, but poor, response when:

- The problem is larger than you thought.
- The problem is one that your planned response can't handle.

Learning too slowly is often caused by ignoring the problem too long, and then being unable to deal with it in its advanced state. We must learn not only *how* to learn, but how to learn *faster.*

2. Manipulating the Symbols of Solution

We have all seen someone deal with a problem in a way we knew was superficial. We know that such approaches won't work, and we aren't surprised when they fail. About twenty years ago, for example, President Gerald Ford attacked the problem of inflation by blaming housewives for spending too much on groceries. No one who had a dictionary containing the term "inflation" could possibly have expected Ford's "Whip Inflation Now" (WIN) campaign to have any effect whatsoever except to anger the women who saw through his outrageous propaganda. Ford was *manipulating the symbols of solution.*

Robert Reich has observed that "Professional education [for managers] is putting progressively more emphasis on the *manipulation of symbols* to the exclusion of other sorts of skills—how to collaborate with others, to work in teams, to speak foreign languages, to solve concrete problems—which are more relevant to the new competitive environment."[4] Managers are often pressured to "Do something!" about some problem or issue. Thus, depending, respectively, upon whether they perceive this as an opportunity that could culminate in either reward or punishment (could win, could lose), or feel the situation has no downside for them personally (could win, can't lose), they may either shift attention and energy away from the real issue to another task they feel they can do well; or they may simply strive to get recognition, press coverage, and rewards by doing something whose value or connection with the original problem can't be assessed. In the first case, the effort may be sincere but mistaken; in the second, the goal of self-aggrandizement is consciously pursued. Symbol manipulation may characterize either case.

Not long ago we saw President Bush answer the problem of the decline of American industry by traveling to Japan to tell its government to stop making it difficult to sell American products there, especially automobiles.[5] Some observers pointed out, however, that Mercedes Benz and BMW are quite successful selling in Japan. Others noticed that Bush's cries of "unfair" came on the heels of the fiftieth anniversary of Pearl Harbor and Japan's "apology" for the attack. Why is the *Japanese* market the key to *U.S.* recovery? Have we no trading partners elsewhere in the world than a tiny island nation no larger than the state of Montana? More profitably perhaps, one could ask why the American consumer so often prefers products made offshore: "Lower price" hasn't been the reason for years. (We shall return to the topic of international trade in later chapters.)

At home, the Bush administration fanned the fires of pseudo-patriotism and racial bias with a "Buy American" campaign, creating schemes giving income tax reductions to those who helped contribute to the profits of certain U.S. companies by buying their products. This propaganda is an example of what Edward Hall terms *extension transference:*

> *a common intellectual maneuver in which the extension is confused with or takes the place of the process extended . . . mistaking the symbol for the thing symbolized while endowing the symbol with properties it does not possess. Worshipping idols, common to all cultures, represents one of the earliest examples of the ET [extension transference] factor. In the Bible, we see this when men are directed to give up the worship of "graven images."*[6]

Buying American products is an *effect* that will be *caused* whenever American consumers consider American products more desirable than foreign ones. By attempting to stage the effect, the propagandists (and their sponsors) hoped to simulate the cause, but without making the products any more desirable—thus the extension transference.

Another response to the problem, hatched during the Reagan[7] administration, is competition for the Malcolm Baldrige National Quality Award. Instead of creating a qualification-based award—one that would be accessible to all who meet the qualifications—the government chose to entrench the paradigm of competition for artificially scarce rewards by offering only six Baldrige Awards per year.[8] Both the judging and the awards ceremonies have become grand theater, with management often willing to spend millions in order to reap the benefits of the publicity and to gain the special seal winners are allowed to display and discuss in their advertising.

An entire industry has sprung up around this competition, offering prep courses on preparing the application and "inside" tips on what impresses the judges. Some consultants offer to write the application themselves, virtually guaranteeing success. The winners are assured of copious opportunities to brag publicly of their "success"; in fact, doing so is considered a *responsibility* of Baldrige winners. In addition to publicizing the receipt of the award, "recipients are required to share information on their successful quality strategies with other U.S. organizations."[9]

Companies, and even city and state governments, rapidly appropriated the Baldrige scheme and began their own internal competitions for shadow awards as preparation for the main event. Managers within these organizations outdo one another in creating presentations that portray them as being well along the course of aggressively pursuing quality with victory certain in the near future.

Each of these responses is an example of manipulating the symbols of solution instead of changing anything in the real world. Of course, Baldrige competitors may actually *make* some improvements, but the rules of the game make few of the necessary distinctions between symbols and reality.[10] Some of these symbols currently being manipulated are:

world class	total quality management[12]
assuring success	excellence
zero defects[11]	success story
empowerment	voice of the customer
participative management	benchmarking
self-managing work teams	re-engineering
leadership	six sigma

We live in an age of symbol manipulation, by television, popular music, press releases, movies, CCTV, fax, VCR, newspapers, and so on. We hear the term "virtual reality" and wonder what it means; perhaps we really don't want to know the answer. The longer the symbols are manipulated the more they are accepted as reality.

President Ford's disingenuous WIN campaign had no effect on inflation because it merely manipulated the symbol of a visible scapegoat: the housewife, who was downstream of the system of causes of inflation. Had the government, instead, stopped creating purchasing media (money) in excess of the value at current price levels of newly produced goods available for purchase in the markets—which is what inflation is—the *effects* of inflation, the continuing rise in prices, could have been altered.

Bush's attempt to manipulate the symbol of Japan's "unfairness" was met by counterattacks by Japanese politicians; but nothing was done in the real world by either side to solve or even recognize the real problems of our industrial decline.

As the heat of Baldrige competition mounts, both government and corporate management find that manipulation of the symbols of progress is far more accessible than actually making progress. The symbols exist in people's minds or in presentations, but they are not aligned with facts. The Baldrige competition is another example of extension transference: in the minds and actions of management the competition itself frequently has taken the place of the improvement process which ought to have been put into effect in order for an organization to qualify for a prize.

Cleverness in symbol manipulation is often confused with cleverness in attacking the problem. Such cleverness is one sign of someone who is skilled at operating *within an existing paradigm,* who has developed a cynicism toward attacks from outside it.

If the paradigm has shifted, as it had in both of the Detroit automotive examples, cleverness helps speed the path toward failure because manipulating symbols has no counterpart in the world of reality where things are happening. Symbol manipulation is a major obstacle to acceptance of Deming's profound changes; it lends the illusion of a concerted attack on a recognized problem, yet often even the problem itself is misunderstood and misstated.

Illusions are sometimes uncovered for what they really are, but often only after great economic loss. Even then, there is no assurance that one illusion won't be replaced by another. As long as virtual reality can be made to appear real, people whose motives are to use the system for their own ends will manipulate the symbols of solution in place of solving problems.

On a given problem there is no assurance of breaking out of the symbol-manipulation phase because of its noetic effects: the longer and more successfully we manipulate symbols, the less our reason and intellect can deal with real solutions, and the stronger will be our consensus that we are doing the right thing.[13] If we ever do break out of this phase, it probably happens in one of these ways:

- You come to see that what you've been doing is a substitute for solving the problem rather than a solution. How this happens depends on whether you were deliberately indulging in subterfuge or were just lost.

- You receive some kindly help. The helper may not know the solution, or even be conscious of helping you, but somehow this person gives you a new way of interpreting the problem.

- You learn something new which leads you to a new way of interpreting the problem.

- You somehow get smarter.

Learning is interventionary in nature: It comes from outside, via theory, or from contradictions we observe. Whether it occurs through personal interaction or an experience, learning can come only by invitation. As Deming says, experience without theory teaches nothing. This entire book is an attempt to provide you with theory that will allow you to interpret past and future experience for the purpose of undergoing a series of significant life transformations. We will talk more about theory as we move on. The remainder of this section assumes you have somehow undergone a transformation out of the limiting phase of symbol manipulation.

3. Solving Problems

Solving problems is the initial productive phase of learning. It begins by identifying something as a problem and admitting that a real solution will be needed. So many problem solving approaches are available that we will not attempt to list even some of them; many good books describe them. A characteristic of this phase is that you solve problems without knowing just how you solved them. For this reason, simply solving a problem, although a great step beyond avoiding it and trying to get credit for moving toward the solution, doesn't necessarily prepare you for solving a similar problem again. In fact, each problem tends to be in a class of its own, and when we try to extrapolate our solution technique to other problems we often find it doesn't work.

One reason for this failure is that we so often act intuitively or out of habit, with little thought as to the processes or assumptions we use. Our thoughts are on the problem, not, as Deming says, on the method by which we are trying to solve it. We end up facing some problems many times over and approaching them the same ways each time; and we worry about having to apply these ways to new problems.

Certain characteristics about this phase of learning are worth clustering together:

- This phase is an extension of the intuitive or unconscious problem solving methods adopted in childhood.
- We don't go searching for problems—they find us. And because all systems have variation, complexity, bottlenecks, constraints, and other built-in flaws, finding problems is usually fairly effortless!
- We tend to be deductive: we are *applying* principles rather than searching for new ones. We often deduce causes by looking at effects.[14]
- Often our direct sensory experience is our only evidence of what works and what doesn't.
- We concentrate on the problem itself, not on how we are approaching it or solving it.
- We are fairly uninquisitive about what we don't know. This hinders us from gaining the insight to form the hypotheses—the attempts to reduce the amount of what we don't know—which would lead to new knowledge. Learning is obstructed and delayed. In Deming's terms, we are trying to learn through experience without theory.
- We tend to search for examples of others' solutions to the same problem. Whether discussing problem solving or choosing improvement approaches, Deming is adamant: "Examples teach nothing. One can only learn theory, and revise or extend theory."
- Because our approach is to try whatever works, and because we often rely on deduction, we often fail to differentiate between *symptoms* of problems and the problems themselves.
- The shortcomings above often lead us to ignore process variation or to confuse its special and common causes, resulting in tampering and random adjustments to the process.
- In this phase it is only natural for our solutions to be rather

narrowly focused on the problems at hand, and for us to overlook the effects that our specific solutions have on the system as a whole. What *we* gain, some *other* part of the system loses. Deming calls this suboptimization.

- In terms of the hierarchy of change described in section 7.2, "Understanding Change," our problem solving is *changing the way we do things.*

The approaches to problems given elsewhere in this book, such as in section 10.2, "What to Work On," involve ways of learning about problems, and becoming aware of how we solve them, at the same time we are working on them. This approach allows new knowledge to be created about the problem itself as well as about the process of solving it.

4. Defining Problems

With defining problems, the chief feature of this phase of learning is seeking out problems rather than simply reacting to them. But this amounts to more than a simple change of stance from defense to offense; it means that you have begun to take into your own hands, at least some of the time, the definition of what constitutes a problem.

In the section "What To Work On" we will suggest an approach that deliberately chooses a problem and gives ways of defining it, instead of merely accepting any of those clamoring for attention. An advantage of being the one who defines the problem is the ability to solve it on one's own turf in an active rather than reactive mode.

Some characteristics of this phase, defining problems, are as follows:

- We are moving well away from the defensive way we handled problems in our youth.
- We are beginning to define what it means for something to "*be* a problem."
- We can appreciate variation and statistical control, and rarely engage in tampering with processes.

- Although we are still using deduction, we are beginning to see the need at times to create **new knowledge**, and are using the scientific method to do so when necessary. Compared to the previous phase, learning is occurring more effectively and more frequently.
- Our method is based upon theory plus experience, rather than on experience alone (trial and error).
- We are much more aware of *how* we solve problems, and are using what we've learned previously toward solving today's problems.
- We are becoming increasingly aware of, and concerned with, what we don't know. Empirical evidence, for example, is never complete.
- Occasionally our analysis turns up data that are inconsistent or contradictory. A constraint typical of this phase is that these occurrences usually bring our learning to a halt.
- Our solutions show more consideration of their effects on the system as a whole. At times we suboptimize, but increasingly this is an oversight rather than business as usual. We are developing an active, conscious win-win approach.
- In terms of the hierarchy of change described in section 7.2, "Understanding Change," our new approach to acquiring knowledge (our epistemological framework) could roughly be correlated to *changing our mind-set and theories.*
- Because of these advances in our own learning, others are beginning to seek us out to share in what we have learned.

Using a theory-based approach instead of merely relying on intuition means you can get at the issues that are really important for systemic improvement much more quickly, as you are no longer merely being driven by pressure or by availability of piles of data. You will define and develop the data you need as you go along your chosen path. This method will also tend to indicate the analytical tools you will need to use in place of reliance on non-analytical problem solving techniques.

5. Questioning Problems

At this point problems are being selected rather than accepted, and the scientific method is being used to solve them. New knowledge is being gained. We understand how variation can be induced in a process by the actions of management.[15] But the longer we remain in a given problem area, the more our analyses reveal patterns that are inconclusive, preventing us from gaining the knowledge we need to go deeper into the problem. We become very aware that, as George Box pointed out, both critical events *and* perceptive observers are comparatively rare.

Another mark of this phase of learning is that even for problems that we know exist we begin to question *why* they occur rather than accepting them. This leads us further toward sources of problems and further away from their symptoms.

Scientific method uses existing knowledge, plus new theory, plus experimentation. The theory attempts to predict the results of the experiment. If the experimental data validate the theory, it begins the passage from theory into knowledge; as the number of cases that test the theory increase, it is accepted to greater and greater degrees. Modern science is skeptical about new knowledge; theories are always subject to refutation, even though held for a long time. Einstein's theories of relativity, for example, might be found inadequate even after standing for most of this century if an experiment in some domain which previously could not be tested turned up refuting evidence.

Scientific hypotheses are "conceived of with a view to coordinating the known facts of experience. As such they may vary from time to time as our knowledge is increased or becomes more refined."[16] As Euclid told King Ptolemy in the days of ancient Alexandria, there is no royal road, no short cut to knowledge. We must therefore test these hypotheses, hoping for a great degree of generality in their application, for without generalization there can be no prediction.

Our interest in the following discussion, however, will be in what happens to a theory when new data turn up that are inconsistent with it, or that contradict it. Is the theory rejected entirely? Is everything reset to zero? Let's consider that for any theory an experiment might have four alternative outcomes:

1. **The results of the experiment are inconclusive.** This not only fails to prove the hypothesis true, but it also fails to disprove it. One reason could be an inaccurate process of measurement; another could be fuzzy expression of the hypothesis itself. In any event, if conclusions are what we are seeking, another, different, experiment will be necessary.

2. **The experiment "proves" the theory.** Put in more scientific language, the data yielded by the experiment fall within the range predicted by the theory, and thus the experimenter "fails to reject it" as a hypothesis.[17] At this point, more tests may be run to determine how large a domain is covered by the theory; others may begin their own experiments to convince themselves whether the new theory accurately describes reality as they see it. If the theory is shown to apply widely in a strategic area of science it may ultimately be called a law, but this process may take generations to occur.[18] Of course, at any time new experiments may find unanticipated problems requiring the theory to be modified.

3. **The experiment "disproves" the theory.** Again, a better way of thinking about this is that the experimental data were quite different from what the new theory or hypothesis predicted; the theory is shown to have no predictive value in the domain in which it made its claims. If no further work is done on the theory, it may fall by the wayside, perhaps being noted as a false lead or dead end for others to avoid.[19] Having this happen to one of your hypotheses may seem a misfortune. However, in his classic book *Science and Hypothesis* Henri Poincaré makes this observation:

The physicist who has just given up one of his hypotheses should, on the contrary, rejoice, for he found an unexpected opportunity of discovery. . . . If it is not verified, it is because there is something unexpected and extraordinary about it, because we are on the point of finding something unknown and new. Has the hypothesis thus rejected been sterile? Far from it. It may even be said that it has rendered more service than a true hypothesis.[20]

4. **The experiment generally "proves" the theory but with exceptions.** Somewhere in the theory's claim-space is a region where the experimental data are contradictory or inconsistent: the theory simply doesn't work in that region. One might be tempted to write this off to inaccurate measurements and assume that no hole exists in the theory, or that it is a special case and someone will figure it out later. To do this, however, is to throw away a major chance to learn, for—assuming the measurements were correct and the inconsistency is real—we are confronting a region we really do not understand which may require quite a different theory to describe it.

In earlier phases of learning, we were used to the second or third experimental outcome, as they divided things cleanly into black or white—the theory either worked or it didn't. The fourth possible outcome we found confusing, as it seems to muddy the water; yet identifying contradictions is key to developing new theories and increased learning. We *must* not cover this region up as though it didn't exist. Discovering the laws that govern it may lead to a great deal of new and deeper knowledge which would be unapproachable any other way. Deliberately choosing this path is the heart of this phase of learning.

We started the discussion of this phase by noting the characteristic of questioning why problems occur. This means moving in the direction of more cosmic issues, although we may also be uncovering lower-level issues, or less complex solutions for them which were not known before. Where contradictions or inconsistencies are noted we will investigate why something would only be a problem within these regions.

The questioning phase has some additional characteristics:

- A bias toward system-optimizing activity is now fundamental and explicit. We routinely weigh the effects we are having on the culture of the organization.[21]

- Teaching others has matured to the point where we are preparing others to teach. Perry Gluckman did this so naturally and effectively using the Socratic method that his students always felt they were having comfortable, cordial dialogues instead of being in class.

- In terms of the hierarchy of change described in section 7.2, "Understanding Change," our new methods for obtaining knowledge, plus our recognition of its limitations, are *changing what we think the universe is.* Our cosmology formerly assumed a clockwork universe, but we are now coming to see a nondeterministic universe beset with variation.

In this phase, then, we find not only better discrimination in our theory-making, but a capacity for second-order thinking about problems—questioning why they occur. Both of these may be very challenging, but in return for meeting each challenge we may gain a substantial increase in knowledge compared to earlier phases.

6. Adopting a Principle of Learning

In this phase, the previous level of learning is adopted as a part of life, to be applied widely in many different areas rather than only at work or just on a certain class of problems. This transition could take a long time. It is aided by the continuous learning process which has been going on in the previous phases, plus any learning occurring in other areas of our lives. The discoveries we are making bring us *joy in learning.*

Several important characteristics mark this phase:

- We recognize the provisional nature of knowledge—we understand that our knowledge is never perfect and is subject to revision at any given time.

- We do not expect new knowledge to jump into our minds without effort or study; we expect to pay a price for it.

- We recognize more of the synergetic nature of knowledge—how what we have learned in one field can, and should, be true in others. Knowledge is not compartmented, even though it is often taught that way.

- Despite—and even because of—these limitations we expect and plan to continually learn new things which will challenge, modify, and deepen our current knowledge.

- We consistently use higher order thinking (examining, and reflecting on, what we are doing).

- As a matter of course we use the scientific method to test and increase our knowledge.

- We understand that plans, procedures, relationships are all assumptions. We don't blame ourselves or others when outcomes differ from predictions or expectations; instead, we seize them as opportunities to learn new knowledge.

- We are intrinsically driven to accept nothing less than holistic solutions, and our system perspective has widened to include not only our immediate group but also our organization and the community of which it is a part. The value of individuals is never diminished or discounted.

- We not only teach explicitly, but by example as well. Others, drawn by the integrity of our principles and behavior, begin to exhibit some of the same characteristics. Deming is just such an archetype, unwavering in fundamental issues but always willing to learn—and skilled in doing so.

- In terms of the hierarchy of change described in the section "Understanding Change," our progression into this phase would inevitably be marked by *changes in our values,* a transformation Deming unremittingly insists on.

Our goal in this chapter has been to describe a continuum of learning that should occur as one grows in understanding of Deming's profound changes. Rather than present a manual of specific personal steps for achieving this growth, we have brought together many of the thematic lines that run throughout this book: the sources and results of complexity; the need for scientific method and a theoretical approach; the strength of paradigms; higher-order thinking; the open-ended and probabilistic nature of knowledge; and the dangers and limitations of intuitive problem-solving approaches.

NOTES

1 Barker's videotapes and books on paradigms are listed in the bibliography.

2 From Hall's book *Beyond Culture*, Anchor Books, 1976, pp. 35 ff.

3 Argyris, long a researcher in motivation and organizational development, argues that managers, while espousing theories of excellence, equity, and empowerment, actually carry out quite different theories leading to defense and cover-up, unconsciously practicing skilled incompetence resulting in counterproductive actions. See his *Overcoming Organizational Defenses: Facilitating Organizational Learning,* Allyn and Bacon, 1990.

4 *The Next American Frontier,* 1983, p. 160.

5 Today the same tableau is being acted out; only the players have changed.

6 *Beyond Culture,* pp. 29–30.

7 Lest we seem to pick only on Republican administrations, we note in passing that similar examples could easily have been drawn from Kennedy's New Frontier, Johnson's Great Society, Roosevelt's New Deal, or many of Clinton's programs.

8 The Deming Prize, to which the Baldridge Award is distantly related, is essentially non-competitive, being awarded to as many companies as meet its qualifications.

9 *1993 Award Criteria for the Malcolm Baldridge National Quality Award*, U.S. Department of Commerce, p. ii.

10 As noisome as is the theater of Baldridge competition, the award's chief flaw lies in the Tayloristic presuppositions underlying the examination criteria applied to the contestants. Even organizations that choose not to compete for the award are beginning to use its criteria for internal assessment, thus spreading Taylorism under the semblance of being near the leading edge of quality improvement.

11 Deming calls zero defects "a totally nonsensical concept." Two of the more important reasons are (1) Its unrealistic built-in binary worldview (in spec, no loss/out of spec, loss); and (2) In-spec product may be of no value to the customer because it doesn't fill his needs. Deming's philosophy of a *loss function* depicting continuous and increasing loss from *any* deviation from the target value, and of continuous *market research* to design products that anticipate customers' needs, supersedes the zero-defects concept.

12 Deming has repeatedly disassociated himself from this symbol. People who ask him about "TQM" almost never define what they mean by the term, and the few who try usually include some elements that either violate his philosophy or reflect little understanding of it. Self-managed work teams, for example, sound very Deming-like—but unless they are *managed as a system* their actions will suboptimize the total system; unless the teams understand variation they will tamper with their processes by listening only to the voice of the customer (specifications, comparisons with averages) instead of first listening to the voice of the process (via control charts).

13 For more insight on how policy and decisions are made this way, see Jerry B. Harvey, *The Abilene Paradox and Other Meditations on Management* (Lexington Books, 1988), and Irving L. Janis, *Groupthink:*

Psychological Studies of Policy Decisions and Fiascoes (Houghton Mifflin, 1982).

14 As Irving Copi notes in his popular book, *Introduction to Logic* (7th ed., MacMillan, 1986), "a deductive argument involves the claim that its premises provide *conclusive grounds* [of proof] . . . . An inductive argument, on the other hand, only involves the claim that its premises provide *some support* for it." A deductive argument made by Taylor or Maslow might be, "Employees are motivated by money, status, or power. Ms. X is an employee. Therefore Ms. X is certainly motivated by money, status, or power." The scientific method as taught by Shewhart and Deming, however, recognizes the tentative and necessarily incomplete nature of knowledge and proceeds by means of inductive arguments.

15 See section 6.5, "How Management Promotes Complexity."

16 *The Evolution of Scientific Thought from Newton to Einstein*, by A. d'Abro, (2nd ed.), Dover Press, 1950, p. 408.

17 Hypotheses can never conclusively be proven. One of the reasons is that experimentally derived data can typically be accounted for by more than one mathematical relationship or formula. The experimenter's "failure to reject" is language carefully chosen to avoid a claim of having proven a hypothesis.

18 Observed on a bumper sticker: "$E=mc^2$—Not just a good idea; *it's the law*." Yet even this is not a constant relationship: as the speed of light (*c*) is measured with increasing accuracy, its "true" value keeps changing. Deming shows this graphically in *Out of the Crisis*.

19 Another interesting property of hypotheses is that, although they may be rejected by the experimenter for lack of experimental evidence, they cannot be *proven* false. A proof, whether of truth or falsity, would have to rely on the truth of certain background assumptions, such as other scientific laws and properties—none of which, as noted, can be proven true. Any proof grounded even partially on unproven assumptions is no proof.

20 Dover Press edition published in 1952. Original English translation published in 1905.

21 See section 9.3, "New Skills For Managers."

From Adam Smith to Deming—An Extension Chain

5.1 Management as an Extension Chain

We introduced the notion of extension chains in Chapter 4. Here is an example of an extension chain. Each of the extensions in this chain brought substantial and irreversible improvements in our ability to communicate.

← wall, changes anyway

1. Spoken language
2. Written language
3. Printing and movable type
4. Word processing
5. Electronic mail, voice mail, fax

(1) We are better able to communicate with a verbal language than without one. (2) The development of written language further extends our ability to communicate and gives it temporal spread. (3) Printing, movable type, typewriters, and now, word processing are each clearly extensions that increase our ability to communicate. (4) In writing this book we chose word processing over typing, writing by hand, or dictation. Anyone whose occupation depends on writing but doesn't use a word processor will have difficulty matching the results of a writer who has one. (5) Few offices today have not been affected by the electronic transmission and storage of voice and printed word. Much of the associated equipment is small enough to be carried by the user or installed in a vehicle.

This is one example of an extension chain that has been developed over time. We have cited only a fraction of the historical extensions. The result of each extension in the chain is that a major function can be done significantly better and easier. Extension chains can be found in countless other areas. The critical point is that we must understand extensions and continue to add to their chains if we are to remain competitive in the market.

The first step is to identify and understand the existing extension chains. From there, it is up to the imagination and the resources of companies and individuals to develop and add new extensions. Here are some general suggestions for creating extensions:

- Find and work on process bottlenecks.
- Get close to the edge of the paradigm[1] so you can ask questions that may reveal inconsistencies and contradictions.
- (Re)organize work so that patterns of repetition become obvious and may lead to automation.
- Look for complexity, which hinders extensions.
- Look for opportunities to increase division and concurrency of labor.
- Don't concentrate exclusively on the immediate area of the problem or opportunity. Create analogies of the situation, and use them as pointers to research and thinking in other areas which might combine with your direct knowledge to produce a breakthrough.

Here is another example of the extension chains we discussed in Chapter 4. It outlines the development of industry and, consequently, of management. The influence that previous thought has had on both Taylor and Deming is obvious.

- Working together (prehistoric)
- Creating products (ancient)
- Trading in materials and supplies (ancient)
- Principle of division of labor (Adam Smith)
- Market system allowing economic calculation (eighteenth century)

Notes for this chapter are on page 87.

- Interchangeable parts (Eli Whitney[2])
- Using systems (Frederick Taylor and Henry Ford)
- Continuous systems improvement (W. Edwards Deming)

This extension chain above has been developed over thousands of years. Quality and productivity, as well as our standard of living, improve with each extension.

Two hundred years ago, Adam Smith noted this extension chain and characterized it in the first two chapters of *The Wealth of Nations* as the driving engine of Western society. Soon after he published his theory, we significantly increased division and concurrency through the development of interchangeable parts. In the same era, a new economic order arose through the efforts of the so-called Manchester School of economics, allowing producers and consumers to coordinate demand and supply effectively through the medium of prices determined in markets relatively unhampered by government control and favor.[3] Soon thereafter, the extension chain was again added to by Taylor's and Ford's use of standardized systems. Taylor also followed up on ideas that were around for at least a century before, especially the ideas of Adam Smith regarding division and concurrency of labor. The contributions of Shewhart and Deming can be viewed as a logical extension to those great contributions of the past.

Most people can understand the chain up to the point of using systems. But American industry has made little progress since the time of Taylor and Ford. Other than technology, the only significant improvements have been better ways to train employees, better ways to select them, and better communication and motivation techniques. Most people are hostile to the idea of continuous improvement because it represents a new cultural pattern. Many do not understand the nature of continuous improvement, which is continuous change in a consistent direction, guided by an aim or purpose. In order to achieve such consistency, people must figure out how to change by continuously gaining new knowledge—and they are not used to doing that. If we are to remain competitive we must progress to the next major link in the in the extension chain by learning how to continuously improve our systems using Deming's profound changes.

The Japanese in 1950 recognized correctly that the change from Taylor's school to Deming's school represented an extension or paradigm shift that they would have to make to remain competitive. Losing the war provided powerful motivation for them to adopt a new paradigm.

Extensions at Livermore

In 1944, scientist Sid Fernback was a student in one of Deming's wartime statistics classes. Thirty years later as a manager at the University of California's Lawrence Livermore National Laboratories, Fernback remembered Deming's principles and hired statistician Perry Gluckman to help improve the reliability of the computing center.

At the time, Livermore had on-site about one-third of the Cray supercomputers in the world and a programming staff of three thousand. Computer outages were occurring at a rate of one every forty-five minutes, causing interruptions to the highly classified defense work in progress. They had suffered this problem for four years. A Pareto analysis of causes indicated that memory failure was the predominant cause of the outages. After documenting the processes used in running and maintaining the computers, they attached a minicomputer to record vital hardware information as each memory failure occurred. This enabled them to start control-charting the time between failures, and to separate special causes from common causes. This in turn resulted in changes to maintenance procedures, such as reducing the amount of dust in the air. They also discovered through experimentation that the computer memory was especially sensitive to ambient temperature, despite the fact that temperature was already maintained within the computer manufacturer's specifications. Further improvements were made to processes based on this new knowledge.[4]

Using Deming's methods at Livermore, they were able to reach the highest level of reliability in the country for that type of computer, significantly improving productivity. Their analyses also enabled them to predict the remaining few outages that did occur before they happened.[5]

A successful company will constantly develop new extensions, and modify and evaluate. The old, all too familiar pattern of resistance or even hostility toward change must go.

NOTES

1 The edge of the paradigm, a phrase coined by Joel Barker, indicates an area where people hold the current paradigm less strongly. In a company this might be a site geographically or organizationally remote from the corporate offices.

2 In his article "Eli Whitney's Other Talent" (*American Heritage*, May/June 1987), Peter Baida suggests that Frenchman Honoré Blanc, who in 1785 was actually making muskets with interchangeable parts, should be credited with this extension instead of Whitney. He cites no less a witness than Thomas Jefferson.

3 A closely related factor that greatly increased the leverage of free markets was the availability of honest media of exchange (gold and silver).

4 Although Taguchi had not yet promulgated his Loss Function, the Livermore example appears to have been an application of his theory.

5 We are indebted for this account to Dr. George Watson, who worked with Perry on these problems.

Understanding the System and Reducing Complexity

6.1 The Lever of Complexity

In 1981, after working with several clients and making profound changes to their companies, Dr. Perry Gluckman began to realize the result was a system that was well organized and easily understood. Waste and delays in the system were significantly reduced, while at the same time he noticed a rather substantial increase in productivity—much more than had been achieved in earlier programs aimed at improving results. The question was how a system that had undergone these particular changes could have as much productivity as they were seeing.

Perry started doing a series of investigations with respect to Japan. He found that the number one effort they made was to *reduce complexity*—understanding and simplifying the system. It's hard to understand how reducing complexity could possibly have as large an effect on productivity as it does; but after observing several clients increase their productivity substantially, Perry realized there were considerable *intangible* benefits—in addition to the more tangible, measurable ones—which came from reducing complexity.[1] Those intangible benefits were what brought about the unexpectedly larger increases in productivity, on top of what could have been expected in gains out of what was worked on directly.

Perry and his clients also were able to observe over time what Perry would refer to as a *nonlinear ripple effect of benefits* across the organization. That is, as the parts of the organization where profound changes were directly made started to see the benefits of their effort, other parts of the organization away from the direct influence of the client area would start seeing improvements in their results and processes as well.

Notes for this chapter begin on page 122.

One example was when a production department worked to improve its testing processes. Upon implementing the production changes, Perry later discovered sales reporting and cost accounting groups whose transaction accuracy and productivity increased. Over a longer time even more departments saw their results improving, without any direct influence from Perry.

Reducing complexity is an important lever in any attempt to seriously improve systems, so the next six sections discuss our discoveries and theories about complexity. They provide some definitions and discuss a variety of intangibles influenced by complexity in systems.

As we go through the discussions on the dimensions of complexity, bear in mind that these factors are not simply additive, but can interact with one another in nonlinear and unknown ways; thus we cannot, for example, simply counteract the factors of system size by applying division of labor and expect total complexity to diminish by a corresponding amount. The overall effect may be less or, as we noted above, more.

Complexity is an important aspect of modern chaos theory, a relatively young field whose contributors from varied fields have been some of the most brilliant scientists of the last century. Here are some insights about complexity from current researcher and author, A. B. Çambel:

1. *Complexity can occur in natural and man-made systems, as well as in social structures.*

2. *Complex dynamical systems may be very large or very small.*

3. *The physical shape may be regular or irregular.*

4. *As a rule the larger the number of the parts of the system, the more likely it is for complexity to occur.*

5. *Complexity can occur in energy-conserving systems, as well as in energy-dissipating systems.*

6. *The system is neither completely deterministic nor completely random, and exhibits both characteristics.*

7. *The causes and effects of the events that the system experiences are not proportional.*

8. *The different parts of complex systems are linked and affect one another in a synergistic manner.*

9. *There is positive or negative feedback.*

10. *The level of complexity depends on the character of the system, its environment and the nature of the interactions among them.*

11. *Complex systems are open in the sense that they can exchange material, energy and information with their surroundings.*

12. *Complex systems tend to undergo irreversible processes [of change].*

13. *Complex systems are dynamic and not in equilibrium; they are like a journey, not a destination, and they may pursue a moving target.*

14. *Many complex systems are not well-behaved and frequently undergo sudden changes that suggest that the fundamental relations that represent them are not differentiable [that is, they cannot be split apart or reduced to smaller components].*[2]

These observations come from an advanced science textbook, but are uncannily accurate in describing the systems we live and work in every day. You will probably recall some of them in our discussions in the next several sections of the book.

Traditionally, the approach to studying a complex system has been reductionist: breaking the system down into successively smaller pieces, and trying to learn about the system by aggregating the knowledge of individual pieces. This is standard practice in business or government. In contrast, Deming's approach is holistic, emphasizing the need for the system to have an aim and to be managed as a system; stressing that system optimization is not brought about by optimizing individual elements; and telling management they must stop putting people and departments in competition with one another.

In the day-to-day application of Deming's philosophy, managers must at times be able to grapple with details in a manner that might be called reductionist, yet always maintain the superordinate, holistic view of the system in order to guide it toward its aim.

6.2 Dimensions of Complexity—System Size

We have found six dimensions to the concept of complexity. We have divided them into two groups of three: The group studied in this section reflects the system size; the other group, studied in the following section, reflects the system impact. Although the following descriptions might tend to create an impression that the dimensions are independent of one another, they have more of a tendency toward interaction—after all, we are describing characteristics of systems.

The descriptions for the system *size* dimensions of complexity are the number of units, the volume of the system, and the density (smallness or largeness) of the system.

Number

Relative to *number*, as an example consider the employees ("units") in an organization and how as their number increases, the complexity grows. In part, this is true simply because more communication "links" are required for all employees to receive information on the processes and state of the organization. Many organizations develop a hierarchy of managerial and departmental levels as they grow in size, but in most cases this merely results in further growth of complexity. (More on this later.)

Similarly, complexity is affected by the number of units in a work batch or lot. More non-value-added handling is required to keep control of the units not being immediately processed, so smaller lot-sizes can help reduce complexity.[3] Similarly, the amount of information required to sustain an activity tends to affect the level of complexity present.

Volume

Complexity also changes relative to the *volume* or space taken up by a product or task environment. Compared to a home garage, a work space the size of a 40,000 square foot warehouse has far more inherent complexity just in workers traveling from task to task, as well as in factors such as climate control and maintenance of the building. This can also be addressed from the standpoint of not having *enough* volume within which to do the work prescribed; if the work is not simplified in some way, then too little volume available can also increase the complexity involved in accomplishing a task. An example of too little volume being available would be when available computer memory is inadequate to manage all files and tasks; the processor becomes overloaded moving data in and out of what memory is available. There is no absolute scale here—what is important is the *relative* volume. In addition, the more that volume is under or over what is required to adequately contain an object or activity, the more complexity will exist.

Density

Similarly, in situations where *density*—the size of something relative to its components—is a key characteristic, the complexity will tend to increase. Whether the subject is relatively large or small, the more tightly packed its components are, the more complexity can be found. Increasingly compact integrated circuits are fine examples of this as more sophisticated processes are required to design and manufacture physically smaller circuits with increasing numbers of semiconductor paths and junctions. And, as with volume, low density may also increase complexity. Consider, for example, the added effort required to maintain consistency and communication when regional sales/service offices are widely scattered and there is a need to get agreement and buy-in to a new marketing strategy quickly.

For years, the traditional business school has assumed that as a company grows bigger and bigger it should get increasingly higher efficiencies. This is not always the case, though, as is illustrated by the recent troubles of General Motors, IBM, and other giant companies. Recently, some efficiency experts have discovered that if you break up a company into parts, which are coherent, self-standing systems, the efficiency of the parts is greater than the whole. We can now see how this result can come about due to the reduction in complexity within the separate units.

Companies will tend to get oversized and they become inefficient against competition because of the increasing complexity that comes with size. However, focusing on *reducing* levels of complexity in all its forms within a corporation even as the corporation expands can lead to substantial *increases* in efficiency. Breaking a company up is just one (rather drastic) alternative to achieving greater efficiency and excellence as growth takes place.

The keen observer or analyst will find a multitude of ways in which complexity creeps into a growing organization. When attention is paid to the chain of causes creating increased complexity, action can be taken to counteract or remove some of those causes. The organization will have a better chance of growing at a faster rate, without having to bear the costs additional complexity could otherwise force upon it.

6.3 Dimensions of Complexity—System Impact

The following are the other complexity dimensions that can be reduced as we learn how to use Deming's profound changes. The level of complexity contained within a system will tend to influence these factors, so these factors will be hints to the analyst that levels of complexity exist within the system. Note again that if we intelligently adopt Deming's profound changes as a company grows larger, we can actually cause the overall complexity of the company to drop. Deming has understood the power of complexity specifically within variation for many years. However, variation isn't the only factor creating complexity that we have to work with.

Process Time

It is very effective to look at process time as having an influence on complexity. For process time, the scale we use is natural to everyone—and we know the direction in which we want it to move. Consider the value of applying the scientific method and reducing process time within a large company (which would tend to have a lot of complexity): as decreased process times tend to produce lower costs and more units, the result would be the large company out-competing a small company without any trouble at all. And, as we learned from David Packard, decreasing cycle time has a direct and proportional effect on the organization's ability to fund its own future growth.

Variation

Variation, whether random, deterministic, or chaotic, is another factor driving complexity. Variation brings about a need for action and tools to attempt to monitor, control, and reduce the variation. It can force an organization to constantly react, as the variation increases and standard procedures cannot accommodate the variation. Special processes have to be in place to handle the waste caused by extreme variation. Those special processes add to the cost of producing the product or service, but realistically add nothing to the value of the output (unless the organization uses the inherent learning to systematically reduce the variation and improve future output). Deming prefers to use variation for determining system complexity. Our experience working in organizations to reduce complexity, though, has been that using variation as an overall indication of the level of complexity is not nearly as easy as using process time.

Context Level

Context level, another dimension of complexity to consider, is drawn from cultural anthropology and the work of Edward Hall. In his words, "The level of context determines everything about the nature of the communication and is the foundation on which all subsequent behavior rests."[4] Context level is an indicator of the level of system complexity in the degree to which information is *implied* (high context) versus *explicitly stated* or used (low context) in a communication, and in the degree to which the participants are parties to that particular context level.

One illustration of context level is the so-called man-machine interface—how you and a computer operating system communicate with one another. On the *low context* end are UNIX and DOS, which uses a blank screen with a blinking cursor and a keyboard. All communication with the system is by means of keyboard-entered commands which follow a strict coding scheme that has no counterpart in real life. If you make a mistake, DOS's response will most likely be the message "Bad Command Or File Name." No help is provided. When you do finally find out what you did wrong, the entire command must be reentered. Many users are reminded of the complexity and frustration of dealing with the court system when using this type of computer. Many have also tried to build a higher-context computer by means of macros and batch files.

Toward the *high context* end are the Macintosh and NeXT operating systems, which present the user with an arrangement of pictograms resembling the surface of a desk, and menus of available system actions to operate on the objects represented. The keyboard-entered command environment is replaced by a system of menus showing what commands are available under any given circumstance, and by implicit operation on the symbols accommodated by a mouse. Responses to the user may be in the form of questions, pictures, or even sound. Much effort has gone into making the system operation closely resemble what is in the physical world of the user. Most users of these latter operating systems feel insulted and estranged when they have to use a DOS-type system, even for a short while. The perspective of differing context levels helps to explain this.

The preceding example notwithstanding, communication is not "bad" or "good" based on relative context level; both levels are appropriate to certain circumstances. High context, for example, is rooted in a shared past, and requires long programming to understand or use. If you don't share this past, you will find communication difficult and will wish you could move an encounter to a lower level of context where communication would be more literal and explicit. What is more important than "high versus low" in determining context level's contribution to complexity is whether the level is appropriate for the participants and the situation.[5]

Context level may rise, for example, when the number of required unit operations in a manufacturing process increases, if there is not a corresponding increase in the work group's explicit knowledge of how to apply those added steps to the process. And as each step introduces greater degrees of variation, the inherent amount of information required to both monitor and sustain the process is greater. In many organizations the context level is so high that the aim of the system—if indeed one exists—is only dimly understood, forcing its members to rely on their own assumptions as guides for action. The correlative of high context is the high level of situated cognition, or tacit knowledge, required in order to perform one's duties: this is knowledge gained slowly through extended experience with colleagues, not by reading process documentation or the organization chart.[6]

In low context communication, the listener is assumed to know very little and so must be told or shown essentially everything. An example of low context level is the use of a dress pattern, which can be done without knowing any of the theory or procedures used in designing the pattern. (Creating the pattern would, of course, be a higher context level activity.) Using a pattern again and again leads to standardization of a process, thereby reducing its context level and lowering the level of complexity. (One can clearly see how training—where explicit knowledge can be passed on and standard procedures shared with all the members of a group—has a role to play in reducing complexity.) Total standardization, however, involves selection of steps and procedures based on some data that will be inherently unknown and unknowable; thus, some complexity always remains, even after standardizing.

Our way of describing complexity is not necessarily unique; others could be used.[7] Thinking about complexity in the above way is like locating a point on the earth by means of a system of coordinates: As long as we accurately locate the point, the particular coordinate system used doesn't matter. Perry's investigations do, however, strongly indicate that six dimensions of complexity will exist no matter what approach to describing the complexity is used; the above six dimensions have proved to be useful.

To reiterate an earlier comment, these six dimensions not only have their individual influences, but will also tend to combine and interact to affect system complexity further. The concept has been likened to a tetrahedron—a four-sided solid with six edges somewhat resembling a pyramid. Each edge represents one indicator of complexity, and the volume enclosed is the level of complexity in the system. As you act on one dimension of complexity, its respective edge shortens or lengthens. The other edges will have some tendency to adjust themselves correspondingly in order to maintain the level of complexity. This representation helps a practitioner appreciate how, when acting on any one factor influencing complexity, one must simultaneously monitor the other factors and check that they are not counteracting the desired improvement. Likewise, if one factor of complexity is observed to grow, the observer may have to look at the others for solutions to keep the overall level of complexity from growing.

One should keep in mind that this is a limited and deterministic analogy to the concept we are trying to describe—in the real world, reducing a dimension of complexity may actually increase the level of complexity in the system overall. We offer the general concept as another way to look at opportunities for improvement of systems. We also hope students of Deming will try it out in their own situations to discover how it may be of use to them.

6.4 Results of Reducing Complexity

Even though we don't always have strong direct connections with results about complexity, we get outstanding indirect results over and over again.

"What's the ROI (return on investment)? What does detecting and reducing complexity do for us?" These are questions often directed at change agents. In Perry Gluckman's consulting practice in the electronics and chemical industries, he saw direct results amounting to reducing cycle time by 80–90%.[8] But just as important, these improvements also sparked indirect effects in other parts of the clients' organizations which could not be directly measured. These indirect effects were always for the better, but could not always be attributed to specific direct actions taken on the process. This "ripple effect" of reducing complexity benefitted the whole company even though it was resisted at times by people whose jobs depended upon handling, or even increasing, the complexity. The overall result was one of being able to do more with less effort. *Output could grow without needing to invest in more capital or human resources.* The rewards could grow without the risk of further investment. The leverage which comes as a result of reducing complexity should be apparent at this point.

Another benefit of reducing complexity is that progressively one can more easily identify problems and solve them. One way this comes about is through the evolution of an environment more open to (1) evaluation of the current processes in effect and (2) change for the better. As barriers to change and progress are lowered, more information flows that can identify and resolve the problems that are exposed. Similarly, as the amount of implied information needed falls, the level of complexity will also be likely to drop. Reducing complexity usually creates the opportunity to give more attention to the existing problems because less time is wasted on trying to figure out *who* caused the problem and more attention is given to *what* in the process caused the problem. As processes are better understood and problems are solved—with less energy going into defending egos and territory—more learning occurs; theories and knowledge about the underlying processes grow, and more creativity goes into looking at what could be done to improve the system further. That creativity brings about more opportunity for generating "breakthrough" ideas for improving processes. The changes made start coming at a faster rate, taking the organization beyond its tradition-bound peers into levels of excellence not previously obtained. Yet, lest one be too quick to think the job will soon be done, remember that the systems we deal with are dynamic, so new problems will very probably be uncovered which were previously unknown due to the layers of past problems hiding them.

Home runs are serendipitous opportunities that can come from reducing complexity. An example would be the extremely fast setup time for automotive dies in a factory which resulted from reducing complexity in the setup process and elsewhere. The effort to accomplish this might often be resisted by engineers and managers who think there is no ROI in reducing setup time beyond a certain point. But this is similar to the power of a home run in baseball: Once the ball is hit, nothing can be done within the existing system to stop it.

Whereas Western management always demand an ROI for each process change, Japanese management realize that reducing complexity eventually pays great returns. When the transistor was invented, AT&T used it only in its switching systems. In Japan, although there was no demand at the time for smaller radios, Sony's engineers started applying the transistor to radio technology on the basis that the transistor reduced complexity. Today all radios, small and large, use transistors and their extension, integrated circuits. By consciously breaking the paradigm of vacuum-tube thinking, Sony reduced complexity and created an entire new market. Reducing complexity *always* pays off.[9]

Another example of the power of reducing complexity is in the making of auto body parts. The Japanese devoted a great deal of effort to learning how to change the stamping dies quickly, and eventually they achieved the ability to make even a single unit economically. This paid off by providing opportunities for just-in-time (JIT) manufacturing, as well as the ability to iterate product designs rapidly.[10] In terms of their being able to compete, this was a home run. One fundamental reason reducing complexity is such a powerful approach is that for a finite expenditure an infinite gain is reaped. Once reduced, complexity will tend to stay gone[11]—allowing the system to continue pumping out advantages.

The practitioner should keep in mind the danger of simply replacing one type of complexity with another, as in cases where field sales put cumbersome order administration chores on the factory or vice versa. Such an approach may make sense in the short term to match work with resources available, but management must look at the system overall and ask for the long term how the level of complexity can be reduced—as opposed to merely shifted—across the total organizational system. (In this case, possible reductions could be made through changes in product structure or by trimming available options to a smaller number.)

One should not assume management will always welcome recommendations to decrease complexity. The realization that something contributes to complexity will often be a surprise, and surprises are rarely welcomed.

> *Wouldn't we all welcome more laughter in the halls of management? I would be excited to encounter people delighted by surprises instead of the ones I now meet who are scared to death of them. Were we to become truly good scientists of our craft, we would seek out surprises, relishing the unpredictable when it finally decided to reveal itself to us.[12]*

Progress also, of course, depends upon what projects you select for reducing complexity. What you select depends on your knowledge of your business. Refer to section 10.2, "What To Work On," where we describe a systematic method for reducing complexity.

In these two sections on complexity, most of the discussion has been at the level of business processes and products. Deming emphasizes that unless a transformation[13] also occurs in the prevailing style of

interaction between people, teams, and divisions, gains made in production areas will be ragged and short-lived. We will learn further
about these vital interrelationships in the next section, "How
Management Promotes Complexity," and in sections 6.7 ("Lessons of
the Red Bead Demonstration") and 10.1 ("Types of Systems Within an
Organization").

6.5 How Management Promotes Complexity

Management is under constant scrutiny and pressure by stock
market analysts, institutional stockholders, major creditors, and often
the press: each of these blocs or factions wants *results*, wants them fast,
and wants them at the "bottom line," whatever that may mean. This
urgency is rapidly taken up by management who are hypersensitive to
such demands so that they may continue their movement up the career
ladder or even just keep their present jobs. As a result the standard
mode of Western management is an obsession with outcomes (lest they
fall short) and with self-interest (the preservation of career and power).
Even worse for the organization and for society as a whole, the career-
critical outcomes in this perverted system are necessarily very short
term.

One of the most persistent underlying beliefs on the part of these
demanding constituencies is that of the *linear growth model*: a belief in
the ability of a system to continue previous patterns of accomplishment, however briefly such patterns occurred and no matter what
changes will occur in the future.[14] Such thinking encourages the expectation of future outcomes based on previous ones without recognition
of the variability inherent in all observed phenomena. A parent is dis-
appointed when the child gets a B in some subject after having received
a string of As and starts offering rewards for a return to higher grades.
A company has had several years of 10% growth, and its stockholders
demand another such year and another from top management.
Essentially the same thing is going on when the demanded increase (or

decrease) is fixed arbitrarily, without reference to the past; in this case, the demanding constituency is attempting to *create* the linear growth rate by mandate.[15]

No wonder then that managers constantly make demands upon their organizational and production systems that exceed their capacities. When the axe may fall if the quarterly figures don't look good, the temptation to distort these systems (to force them to deliver the demanded results using the traditional expedients of MBO, incentives, competition, and fear)—or to distort the figures themselves—is often irresistible. The fact that this dilemma is needless and destructive, and that another option—improving the system—would help the company, is usually unnoticed or rejected as impractical.

In electing to distort the system instead of improving it, managers force the system to shoulder the heavy burden of increased complexity (see table 4). Each time this is done, the system becomes *less* capable, guaranteeing an unending round of system distortions with necessarily increased levels of complexity as management sets ever-higher levels of performance and exhorts people to meet them. The effects of increased complexity are often subtle and diffused in the organization, making it difficult to trace them back to any specific management action; thus managers who undertake the transformation to Deming's philosophy are typically appalled when they learn the extent to which their own actions have been promoting complexity in the system.[16]

---

1. *Nonlinear Flow of Work.* Imposing policy constraints that disrupt flow

2. *Working Around Missing Resources.* Focusing on goals instead of improvement

3. *Tampering by Over-Adjustment.* Changing the system without understanding it

4. *Promoting Factionalization.* Suboptimizing the system by making its members compete with one another

5. *Ineffective Communication.* Not defining the aim of the system or terms

---

Table 4. How Management Promotes Complexity

This section examines some of the things management does to promote complexity. Their effects may involve any of the dimensions discussed earlier in this chapter. A giant step forward would be for *managers to admit that they promote complexity* through these and other actions.

1. Nonlinear Flow of Work

Hardly a day passes when one fails to observe work being planned and executed in nonlinear ways. In the morning, a group starts working at the tasks left at the end of the previous day. By noon, though, management have decreed the current task must be put aside in order that a rush order can be shipped to impress a new customer with the proclaimed responsiveness of the organization. The next day, another management-imposed interruption causes a key step in the process to be overlooked but found just prior to customer delivery, forcing two days' work to be scrapped. In this example, by not establishing a set of organizational priorities and by reacting to every new opportunity to make a quick buck or close a deal of some sort, managers disrupt the flow of work and cause needless errors and economic loss. The complexity created induces variation into the system, bringing about immediate problems as well as other problems further along in the process.

Nonlinearity need not be physical; it can be temporal: many organizations ship 80% or more of their output in the last few days of the month or quarter. This happens not because efficiency suddenly increases at the eleventh hour, but because quotas imposed by upper management cause operational management to violate the interests of every constituency in the system, releasing orders and shipping out goods ahead of schedule—goods not ordered, incomplete orders, and defective goods—in order to make their targets. When many of these goods return to the seller, they form another kind of nonlinear flow—in the wrong direction.

Organizational policy, not inattention or lack of care on the part of the workers, is the most significant cause of nonlinear work flow.

2. Working Around Missing Resources

Tim Fuller has written an excellent article about the problems and waste brought about in trying to work around missing resources.[17] Because of a circuit board assembly line having to work around queues of product waiting for missing components, productivity was less than one-half what was attained by eliminating back-orders; workmanship defects were reduced by orders of magnitude as the complexity of working around the missing parts was eliminated. His examples are in the production and order management worlds, but the concept is equally applicable to any number of other situations.

Parts are not the only resource that can be missing: if specifications are not agreed upon, a design team cannot provide a new product in a timely way. They attempt to work on parts of the project, but run a greater risk of competing for common tools and are more likely to rework designs as the specifications are defined "on the fly." Specifications are just one instance of a class of resources that could be called *information*, the lack of which will increase complexity. Another resource that is often missing is sufficient time in which to do a good job. The "workaround" is usually to omit steps in the process, which creates the complexity of finding, fixing, and tracking the results of the missing or poorly done steps.

For a variety of reasons—fear of productivity measures dropping, managers' concern that having fewer direct reports will reduce power and prestige, a lack of information on customer priorities, and so on—people and managers are more comfortable keeping busy by working on whatever they can versus focusing on fixing flaws in the system that keep them from being able to work on the right tasks.

3. Tampering by Over-Adjustment

Tampering through over-adjustment is occurring whenever management ignore process variability or fail to appreciate the impact their action will have on the system as a whole. Workers can be driven to making random adjustments when yields are low and when a lack of process orientation is caused by management emphasis on quotas and goals. Specifically, *tampering* is any action taken that is directed toward adjusting the *output* of the process or system without taking any action to work on the *causes of variation affecting it*. Deming uses a clever exercise called the Funnel Experiment, explained in this chapter, to provide insight as to the dramatic negative effects that can come about as a result of tampering.

4. Promoting Factionalization

Factionalization was a term Perry frequently used as a label where he found individuals or groups put into we-they (win–lose) situations, forced into factions whose interests opposed one another. As discussed earlier, traditional management typically pits individuals or groups against one another, often congratulating themselves on how they are contributing to the "health" of the organization. The current euphemism for this damaging practice is "contention management."

The competition for salary budget promotes factionalization. Putting engineers by the windows and clerks in the center of the building does, too. Two design teams may be assigned responsibility to develop the same product, with the intent that the internal competition will produce a superior product; the result is more factionalization, plus the economic loss of time, materials, space, and competitive position. Examples are everywhere; the result is an increase of complexity as barriers arise to sustain the artificial separation of the organization's members. As Deming stresses, one cannot pursue optimization of the system by optimizing its components separately. The organization must be, as Brian Joiner terms it, "all one team."

Wherever factionalization has been instituted, pursuing the aim of the organization (if indeed it is even known) will require additional effort and will most likely involve significant waste.

5. Ineffective Communication

Ineffective communication is yet another way in which complexity is promoted by management. Without clear and effective communication, instructions cannot be followed, suppliers and customers cannot understand one another's needs and desires, and well-intended effort turns into useless output. Deming considers business almost impossible to conduct in a rational manner in the absence of operational definitions, which avoid leaving seemingly obvious terms such as "on time," "average," "white," "complete," or "clean" to the assumptions of the various parties.

With any degree of confusion about what is to be done, complexity again will be present as individuals attempt in their separate ways to make order out of what little information they do have to work with. Besides the passive lack of good communication, there can also be intentional corruption of data for the sake of looking good or avoiding punishment.

In one large company top management, parroting a term made popular by Motorola, demanded that "six sigma quality" be achieved in all products by a certain date. Shortly thereafter, local management began to insist on zero defect levels, announcing this as a step toward "six sigma." When a few people observed that zero defects is a much more difficult target to reach than six sigma, and thus could hardly be a step along the way towards it, they were given blank stares by both workers and management alike. This was an example of ineffective communication—of management's sloppy thinking and slipshod use of technical terms—fostered by upper management who handed out "attaboys" to their subordinates for so aggressively chasing the goals they had so cavalierly set.

Management also increase complexity when they misrepresent as linear the nonlinear nature of the system or any of its elements. This is done by means of reports to higher management, as well as through goal-setting and through demands made upon subordinates or other parts of the system.

When management reorganize, communication among the parts of the system becomes more difficult—even more so if the new formal organization is more elaborate or compromises the informal system.[18] Every time a new layer of approval is required, communication is impeded; information undergoes successive compaction and interpretation at each level until it is bleached of most of its former meaning.

A larger spectrum of offenses could be described, but probably the most damaging to the organization is management not creating, communicating, and maintaining an aim for the system that is clear to all—not having as Deming would say, a constancy of purpose.

All of Deming's original Fourteen Points as well as his system of profound knowledge can be thought of as sharing, among other goals, the reduction of complexity in systems.

Having studied how management promotes complexity, it is interesting to note that many companies today are trying to sidestep their complexity problems by contracting for workers or production by suppliers instead of solving these problems within the company. Instead of heeding David Packard's advice on how fast they can *grow* their companies, management often seek the fastest ways to *shrink* them. In addition to the usual reasons of reducing salaries and benefits, "outsourcing" is often justified by the assumption that it's no longer worthwhile to maintain one's own production capacity because Company X can already make the product at a lower price. Using this logic to change from producing things in a company's chosen product market segment to buying them from suppliers is a fallacy, however, as the company may lose its ability to compete in other areas. The immediate victims of the fallacy of "downsizing" are the employees, who are typically blamed for the problems created by management.

This does not mean that subcontracting is inherently bad. Outsourcing can make sense when *system* complexity is reduced as a result—and when strategic competitive advantage is not lost. The decision must be made carefully. As an example, a company may have very high variation in its monthly forecast submitted to planning. As a result, it takes more planners, inventory, and production personnel—and therefore higher costs—to assure the delivery of product. Looking at direct costs, management decides to subcontract production. It is soon found, though, that the subcontractor cannot make a profit: the same poor forecast causes frequent changes in schedule, high overtime, and unexpected increases in needs for inventory storage. In this case, the decision to subcontract does not solve the root problem—the complexity is merely transferred to the supplier. Again, complexity, cycle time, and *total* cost must *all* be considered in the decision to go to outside.[19]

In addition to the ways described above, management further promotes complexity in these ways:

1. Building data collection, cost accounting and reporting systems that are Taylor-based—which ignore variation, especially outside the area where the data are collected

2. Blaming workers for problems of the (management-controlled) system

3. Creating systems that are difficult to understand or study

4. Working around problems instead of correcting their causes

5. Suboptimizing the whole by optimizing parts ("Run your department as though it were a company.")

6. Withholding information to maintain advantage:
   - of managers over workers
   - in competition among peer managers
   - with suppliers and customers

7. Promoting nonlinear flow of information, through:
   - Technically inadequate communication systems
   - Tradition-bound beliefs about where information should reside in the organization

- Holding back information in order to compete with other elements of the organization

8. Maintaining a hierarchical form of organization, which impedes communication and can lead to competing cultures as well as turf battles

9. Frequent reorganizations or shifts of responsibility

10. Requiring many levels of management to be involved in making a decision

11. Relying on inspection to ensure quality (as Deming says, no one has the job of making the product right), leading to adversarial relations and decline in pride of workmanship

12. Misrepresenting the degree or extent of complexity itself in the organization

13. Using, or allowing, staff (as opposed to line) organizations to make policy[20]

6.6 The Funnel Experiment

One of Deming's most effective illustrations of how management promotes complexity through tampering is through the use of a mechanical procedure called the Funnel Experiment. In it, a marble is repeatedly dropped through a funnel held several inches above a target (an X drawn in the middle of a large sheet of paper) on a table.

The object of the experiment is for the marble to come to rest on the target on each drop of the marble, establishing a pattern of hits or near-misses. Its application to management is obvious. Due to variations in the surface of the table, variation in the shape of the marble, and a host of factors unknown and unknowable, the marble rarely comes to rest on the target. One response might be to alter the position of the funnel after each drop, in the hopes of making the next marble land nearer the target; another would be to leave it as it is. The question for management: which to do?

Deming illustrates four improvement strategies that management might adopt to reach the goal, although others could be imagined. More importantly, he then shows the effects of each strategy on fifty successive drops of the marble, in terms of the changes in variation they cause. We'll take each in the same sequence he uses.

1. Leave the Funnel Fixed

The obvious result of this strategy is variation: Not only does the marble rarely hit the target, but it rarely lands in the same place twice, and never predictably.

As we noted in the section 2.3, "Taylorism and Neo-Taylorism," most systems do behave in random patterns. This establishes the funnel/marble/tabletop system as a real-life system, not just a game. Also, in real life management can rarely resist the urge to "do something"—to tamper with a system that isn't delivering what they want.[21] The other three cases of the Funnel Experiment model these tampering strategies.

2. Move the Funnel from Its Last Position to Compensate

This strategy is complex to carry out, as it requires us to remember both the previous position of the funnel and the position where the marble came to rest on that drop. We then move the funnel in a vector based on the difference between the marble's previous position and the target. The strategy's obvious purpose is to compensate for the behavior of the system observed in the previous turn, and to use the knowledge gained in that one drop in hopes of bringing the marble closer to the target next time.

As the demonstration proceeds, however, it quickly becomes clear that the marble continues to miss the target. In addition, this strategy, adopted to reduce variation and produce a tighter grouping, substantially *increases* variation instead.[22] The cause is that we are trying to compensate for randomness! Some other examples of this kind of improvement strategy might clarify:

- Tuning a station on a dial-type radio

- Adjusting a machine as it is running
- Adjusting a rifle's sights after each shot
- Reacting to a single customer's complaint (for example, with a change of policy)
- Taking immediate action on defects or violations
- Imposing a national 55-mph speed limit and CAFE gas-mileage requirements after the so-called "oil crisis" of 1972–1973

3. Move the Funnel from the Target to Compensate

In this strategy we have given up on getting the marble to land on the target—improvement is no longer the goal. We are just trying to keep its overall performance in "control." On each turn the funnel is moved to a point exactly opposite the previous drop, ignoring the funnel's previous point of aim. If on the last drop the marble landed two inches away from the target at three o'clock, the funnel is now aimed at a point two inches away from the target at nine o'clock.

Before long, however, the similarly shaped distributions of the previous two rules have given way to a curious bow-tie shaped pattern that keeps getting wider and wider off the mark. Scientists in the audience start thinking in terms of resonance and oscillation as the system rapidly explodes, going off the tabletop and, in Deming's words, "into the Milky Way."[23]

Examples of this phenomenon can be taken from life:

- Escalation of bets in a "hot" poker hand
- Short-term "contracyclical" manipulation of credit, money supply, and public works expenditures by the Government trying to cancel out the "business cycle" but actually causing it to occur[24]
- Escalation of confrontational behavior as anger and frustration increase (this was been parodied hilariously by Laurel and Hardy, and other comedy duos)
- Escalation of barriers to trade between nations

- Sending the children to bed early tonight because they stayed up late last night
- Trying to exceed the customer's expectations this month to make up for bad service last month
- Pulling in orders from next month to compensate for cancellations this month

4. Set the Funnel Over Its Last Position

This is another "consistency" strategy; the goal of hitting the target has been abandoned. We are just trying to contain, or perhaps reduce, our losses. Wherever the marble lands, the funnel is simply aimed at that position for the next drop. Deming compares this strategy to the actions of a drunk: staggering along aimlessly until he falls down or runs into something, and then starting off in a new direction without reference to where he was going before. The successive application of random forces results in a random walk.

This rule rapidly destabilizes the system, and it also explodes off the tabletop, only this time in a non-symmetrical pattern.

Examples are immediately recognizable:
- Marking the next board to be cut by using the board previously cut
- Worker training worker (on the job training)
- Hanging wallpaper by using the last piece hung as a guide
- Fitting a gauge to a part to be measured
- Taking a sample from the last batch (as with paint) to be used as the measurement standard for the next batch

Of these four strategies or rules, obviously the first gives better results than the others, yet it yields no improvement. The *best* strategy—Deming's profound changes—we have deliberately not mentioned until last:

1. *Every system has variation; hence, the information needed to create optimum systems is unknown and unknowable.*
2. *Using the scientific method we learn what's unknown but knowable faster.*

> *3. By observing the operation of the system, built-in flaws can*
> *be detected and isolated.*
> *4. Reduce complexity and lower entropy by removing the built-*
> *in flaws.*

By applying this method we could continually learn more about
the system and improve it—never optimum, but always improving. In
the Funnel Experiment this would be shown as ever-smaller groupings
around the target.

Although it lacks the human drama of Deming's Red Bead
Demonstration, which we'll consider in the following section, his
Funnel Experiment is such an unusual model that one may fail to draw
the strong lessons for management practice. Many people become
involved in the various rules, wondering if there might be others
Deming has missed.

The term *tampering* often connotes criminal behavior: because
some criminals have from time to time tampered with the contents of
medicine containers, we now have many types of tamper-resistant containers. Deming, however, uses *tampering* to describe what people do
when they try to improve a process without bothering with the effort
and methods required to learn anything about that process.

Although we can't altogether rule out criminal motives, most
process tampering is done with the best of intent—to "do something"
to make the system deliver what management wants of it. Rules 2, 3,
and 4 of the Funnel Experiment illustrate the fallacy—and harmful
effects—of taking action on each cause as though it were a special
cause. Deming again and again warns us: "People doing their best,
putting forth their best efforts, are killing the country—doing the
wrong thing [tampering]." One reason for this is that tampering
always promotes complexity (randomness, variation), and increased
complexity makes processes less and less predictable. Yet one of management's greatest responsibilities is prediction—not with deterministic certainty, but reflecting a rational *degree of belief* derived upon
analyzed past performance.

Deming's profound changes tell us that to improve a system we have to gain new knowledge about it by using the scientific method, and that the system has to be in statistical control before we can have any convincing degree of belief in predicting its future performance.

6.7 Lessons of the Red Bead Demonstration

Deming's famous Red Bead Demonstration has become a hallmark of his four-day seminar.[25] He uses it to demonstrate how the outmoded paradigm underlying current American managerial techniques makes quality improvement nearly impossible. The paradigm is of course neo-Taylorism. Deming plays the part of a supervisor who, in proper Tayloristic fashion, selects six "willing workers" and trains them to carry out a fixed, rigid procedure.

The workers are to use a special paddle with fifty depressions in it to do their days' work, which is the production of fifty beads. The paddle is dipped into a bowl containing about four thousand red beads and white beads mixed together, and is raised up with fifty beads. The customer requires only white beads—no red ones. Deming tells the workers he will allow up to three red beads in a workload, and warns them that anyone who produces more than three will be put on probation, with more dire consequences for anyone not sufficiently moved by probation to meet the team goals.

What unfolds is a system that never meets its goals, because those goals fail to recognize the variation inherent in the system—they require of it something it cannot give. Yet if any improvements are to be made in that system, Foreman Deming makes it clear that their source will not be the ideas of the Willing Workers, who have been admonished to silence as though they were little different from machines. To keep from closing the plant, the management tamper with the system yet again by ranking the workers and "downsizing," with predictable results: because costs continue to overrun revenue, the plant is closed and the remaining Willing Workers are dismissed.

People who witness or participate in the bead demonstration see the devastating nature of paradigm paralysis, the disease that is keeping, and will continue to keep, American industry from recovery.

Deming used the bead demonstration for years to demonstrate the concepts of randomness and statistical control. He now uses it to bring out many additional aspects of his profound changes.

In terms of what is shown in the sections 6.2 and 6.3, "Dimensions of Complexity," and 6.5, "How Management Promotes Complexity," let us examine some of the lessons of this demonstration.

Variation or randomness is a major dimension of complexity. Randomness is present in *every* system, not just the artificial world of the red beads. Even processes in statistical control can be expected to yield different results each time they are run. Our inherent inability to eliminate this variation completely, or even understand it fully, leads to Profound Change 1: **Every system has variation; hence the information needed to create optimum systems is unknown and unknowable.** Yet in the Red Bead Demonstration, Foreman Deming and his management are amazed by the constant variation in number of unacceptable red beads per workload, and assume it is due to a lack of motivation on the part of the workers.

We know that the other three profound changes offer a way to deal with the variation problem. They can be summarized as *learning about the system and using that knowledge to reduce complexity, repeatedly*. This is management's only hope of improving quality, yet there is no evidence they ever consider it.

During the bead demonstration Deming chides, questions, berates, rewards, punishes, warns, stimulates, incites, goads, and otherwise tries to motivate his team to meet the goals management have set. Yet he fails utterly to change the outcome. Why? Because the workers are utterly powerless to change the system into which they were hired. This system abounds with complexity, such as the rigid sequence of motions to be used in getting the fifty beads from the bowl onto the paddle. Management were unwilling to admit, investigate, or remove this complexity. Nor is the complexity limited to the obvious; it also includes much more fundamental issues.

1. Working Around Missing Resources

- The missing resource here was incoming material consisting entirely of white beads. Two typical workarounds for defective materials are the Acceptable Quality Limit (AQL) or Lot Tolerance Percent Defective (LTPD), both of which deal with *limiting* defects to a set percentage, usually through inspection. This is the system implied in the demonstration when the foreman announces that it is acceptable to produce up to three red beads per lot of fifty; even though the customer wants all white beads, the producer has unilaterally decided that up to 6% red beads is acceptable. AQLs and LTPDs concentrate on minimizing the cost and risk of inspection; Deming's profound changes aim at system-wide improvement—and the system clearly extends beyond even the company to include its customers and suppliers.

- Nearly one-third of the work force was devoted to inspection of finished goods. This increased cycle time, complexity, and the cost of product. Willing workers whom the foreman has put on probation have been observed—even in the context of this *classroom demonstration!*—trying to get the inspectors to falsify their counts so that they won't lose their "jobs." We will shortly have more to say about these effects of the system.

Missing resources increase complexity because of the added steps required to work around, or make up for, them. The result is a system larger than needed to do the work.

2. Ineffective Communication

- Workers were forbidden to discuss, improve, or vary the process, or to communicate problems to management.

Ineffective communication increases complexity because learning and knowledge are delayed or shut off completely, decreasing the effectiveness of division of labor as well as improvements to the system.

3. Encouraging Tampering by Over-Adjustment

- Management tried to get good performance by offering the workers rewards and punishments. This is tampering because the process is in statistical control and the workers are not the cause of the problem; even if they were, the reward and punishment schemes are self-defeating, as we shall see later.
- Management showed their ignorance of variation by firing the "below-average" half of the team. Because of variation, there was in fact no way to determine who the below-average workers really were. Thus the only effect was to increase fear among the remaining workers.
- Management were unwilling to learn about the system in order to improve it, an attitude typical of today's neo-Taylorism. The only "improvement" they tried was firing half the workers, which halved the team's productive capacity and didn't improve quality at all. The workers' performance was determined entirely by the system, yet all of management's attention was directed to the workers, and none to the system.

Complexity and variation are increased whenever management change the system without first learning about it. In the Red Bead Demonstration the results were catastrophic (the "company" went out of business).

4. Promoting Factionalization

• Management dealt with the workers individually, not as a team, and encouraged them to compete with one another rather than cooperate.

Factionalization increases complexity by causing people to work for a subset, rather than the whole of an organization. Maximizing the output of individual subsets of a system does not create an optimum system. When people are made to fear, that subset narrows down to one person.

In other words, management rejected *all* of Deming's profound changes.[26] The consequence was the plant closing. The workers, although their hands actually "made" the production lots of beads every day, had nothing to do with the failure of the plant. None of management's motivational techniques, whether using fear or reward, increased quality in any way.

What are the red beads? The red beads in this demonstration are symbols that transcend the visible physical objects which vex both the management and the Willing Workers throughout. In terms of the lessons of the demonstration, the "red beads" include not only the way the production system is set up—materials, methods, machinery—but also the Tayloristic environment in which the decisions and policies of management, and their ill effects upon workers, customers, and the future of the plant itself, are played out. This constant-cause system explains not only the variation from worker to worker, but also the day-to-day variation in the plant.

Sometimes a constant-cause system is not appreciated for what it is until something changes. One of the authors (Delavigne) was assisting Dr. Deming in a seminar when the average number of red beads in a workload for all four days' work fell from the historical average of 9.4, accumulated over a large number of experiments, to 7.3—a precipitous drop. The results as seen in table 5 for the 11 August 1993 experiment may be compared to those of a 16 January 1991 experiment, shown in table 6. The same beads and paddle were used in both experiments. On days 4 and 5 we see a run of eleven consecutive data points below the previous central value of 9.4. Even though none of these fell outside the control limits, eight or more such occurrences can safely be interpreted as a signal that the process has shifted.

What had happened to the process? Three hundred people, plus Foreman Deming, had been watching closely. The only process change Delavigne and fellow facilitator Ron Moen had observed was a new way of mixing up the beads before each worker's turn. The beads were moved back and forth with the paddle, instead of the traditional process of pouring them from vessel to vessel and back. Later Delavigne found out indirectly from one of the Willing Workers that the new mixing procedure made it easy for a worker to move red beads out of the way of the paddle and lower the red bead count; all the Willing Workers, anxious to keep their jobs, had quickly caught on to this during the training session. Yet unless an observer were actually standing near the bead container, no one would see this "cheating." The system had changed, and the voice of the process sent us a signal.

In a normal business environment, management would have been justified in searching for the cause of the shift in the process. Of course, in this case the process shifted for the better, so instead of removing the cause management would try to stabilize the process at the new, lower

Willing Worker	Day 1	Day 2	Day 3	Day 4	All 4	5th Day
Michael	8	6	11	10	35	
Andrew	6	8	7	7	28	4 8
Linda	8	9	6	8	31	
Dwight	3	4	10	8	25	5 5
Paul	12	5	8	7	32	
Thomas	6	5	9	5	25	9 9
All workers	43	37	51	45	176	40
Cum. Avg.	7.2	6.7	7.3	7.3	7.3	xxx

Table 5. The Red Bead Demonstration, San Diego, 11 August 1993

level of defects, and continue improving it with further experimentation using the PDSA cycle.

The bead demonstration is only a display, a teaching vehicle—but hundreds of millions of people live their entire working lives in systems managed in essentially the same way acted out by Foreman Deming in the Red Bead Demonstration. Most Western management still live in a Newtonian, Tayloristic paradigm that is steadily choking their companies' ability to compete and even survive. Deming's profound changes offer the theory of management they need. When will they wake up?

We can summarize these lessons as follows:

- Every system has uncontrollable randomness.
- To reduce randomness, management must change the system by removing the flaws (red beads).
- Employees working within the system have little or no control.
- Motivating employees, whether by fear, praise, or incentives, will not improve productivity or quality.
- Reacting to natural process variation as though it were a signal that the process has changed is tampering and will only make things worse.

Willing Worker	Day 1	Day 2	Day 3	Day 4	All 4	5th Day
Scott	9	11	7	8	35	16 11
Spencer	6	11	11	9	37	8 10
Larry	12	7	5	5	29	6 9
Seri	11	10	13	9	43	
Tim	14	8	9	11	42	
David	4	11	12	12	39	
All workers	56	58	57	54	225	60
Cum. Avg.	9.3	9.5	9.5	9.4	9.4	xxx

Table 6. The Red Bead Demonstration, Newport Beach, 16 January 1991

NOTES

1 The reader is invited to review Peter Senge's concepts of detail complexity and dynamic complexity for further insight into the theories explaining the type of results Perry observed. (The Fifth Discipline: The Art & Practice of The Learning Organization; Doubleday Currency, 1990; pp. 71ff. and 364ff.)

2 From *Applied Chaos Theory—A Paradigm for Complexity*, Academic Press, 1993, pp. 3–4.

3 In manufacturing, lot-sizes as small as 1 are feasible using a pull- or *kanban*-type system, thereby reducing the complexity that would be inherent to handling larger batches.

4 *Beyond Culture*, Anchor Books, 1976, p. 92.

5 A low context computer interface may be desirable when the aim is to restrict use of the system to those trained in its use. Around small children, for example, the Macintosh might be a great temptation to unauthorized play because many of its symbols are already familiar, and it is easy to get the system to "do something." In order to inhibit such temptation, a DOS interface might be chosen. The DOS screen would probably provoke little interest, and even if children played, the chances of actually entering a syntactically correct command are rather small.

6 John Seely Brown, Allan Collins, and Paul Duguid, "Situated Cognition and the Culture of Learning," Institute for Research on Learning Report no. IRL 88-0008 (December 1988).

7 As an example, Al Viswanathan offers a suggestion that the five components of the Ishikawa fishbone (cause and effect) diagram—People, Material, Equipment, Method, Environment—as described by William Scherkenbach, are an alternative method for describing the components of complexity in a system.

8 3Com Corporation's Customer Repair Services provides an actual case of the direct results to be achieved in how the group brought in-house handling of customers' repair orders down from a 20-day average turnaround to 3.5 days through removing bottlenecks and simplifying the administration and handling of product. Indirect effects were observed in other areas, such as finance where billing the customers became easier, and the stockroom where inventory levels could be decreased as fewer "emergencies" occurred which would have otherwise required carrying safety stocks.

9 In his review of our manuscript, Don Gause notes, "It seems to me the big message here is that companies need a lot more *vision* and a lot less *ROI*.

[Management needs to] replace 'metrics' thinking with 'visionary' thinking." We would not encourage companies to completely disregard the financials but to emphasize the point about vision, we could not have said it better!

10 Note the effectiveness of this approach to achieving JIT, as opposed to "installing" a JIT package without taking details like setup time into consideration and merely reducing lot-sizes to very small quantities.

11 Take caution with this statement, though, given how unknown and unknowable causes (along with the dynamic of change constantly going on throughout the rest of the system) will continue to be at work to introduce new complexity over time.

12 Margaret J. Wheatley, *Leadership and the New Science—Learning about Organization from an Orderly Universe*, Berrett-Koehler, 1992; p. 142.

13 A transformation from neo-Taylorism to Deming's philosophy.

14 As we mentioned in section 2.3, organizations are complex dynamical systems that exhibit nonlinear behavior; thus, expectations of linear behavior—as well as of predictable response to any change in input—are fatuous. Writers and explorers in modern chaos theory have recognized this during the last ten years, but it has been inherent in Deming's philosophy all along.

15 The linear growth model has also influenced the teaching of history, where students are often taught to envision one long, linear march of progress. Continuous dialectical progress is even a dogma of Marxism. Yet certain writers over the last several centuries have proposed cyclical or alternating states of society. See Giambattista Vico (1668–1744), *Principles of a New Science About the Common Nature of Nations*; Oswald Spengler (1880–1936), *The Decline of the West*; and Pitirim Sorokin (1889–1968), *Social and Cultural Dynamics*.

16 They are also equally appalled when they discover that the system they affect is much larger than they formerly imagined, and includes not only the whole organization but also its suppliers, customers, and competitors.

17 F. Timothy Fuller, "Eliminating Complexity from Work: Improving Productivity by Enhancing Quality," *National Productivity Review* (Autumn 1985), 4(4), pp. 327–344.

18 The informal system is the system that actually does the work, as opposed to the formal system, which is spoken about by management as though *it* actually describes the way the organization works. Further elaboration is given in section 10.1, "Types of Systems Within an Organization."

19 In addition to the factors discussed above, there is also the issue of core

competencies, that is, those functions the company does so well that they are essential components of its competitiveness and should only rarely be allowed to be done by others. See C. K. Prahalad and Gary Hamel; "The Core Competence of the Corporation," *Harvard Business Review* (May–June 1990), pp. 79–91.

20 Deming provides one extended example of this point in *The New Economics for Business, Government, Education* (M.I.T. Press, 1993), pp. 70–73. Most of us could supply many others.

21 See Professor William Haga's chapters, "The Shiftscheme Response: Do Something!" and "The Firecracker Response: If You Can't Do It Well, Do It Big" in his perceptive book *Haga's Law* (William Morrow, 1980).

22 An often-employed variation of this rule is to adjust the funnel (process) only if the marble lands outside of a circle centered on the target and of diameter d_{spec} creating a "deadband" in which no adjustment takes place. A recent paper demonstrated, however, that the process variance for such a deadband will eventually approach that of no deadband at all—twice the variance yielded by rule 1. Gitlow, Kang, and Kellogg, "Process Tampering: An Analysis of On/Off Deadband Process Controlling," *Quality Engineering*, 5 (1992–1993), pp. 293–310.

23 What is "the Milky Way" in this example? Clearly not literal interstellar space. Deming is referring to behavior viewed as chaotic, unpredictable, no longer amenable to control. In business we might hear management speak about something that "got totally out of hand and had to be shut down."

24 An example is given on p. 199 of Deming's *The New Economics*.

25 The Red Bead Demonstration is also shown and explained in volumes 7 and 8 of *The Deming Library Series*, an educational set of videotapes from Films Incorporated in Chicago, IL (telephone: 800-323-4222).

26 Although not as pertinent to this discussion as the other issues above, it could also be noted that management failed to exploit many obvious opportunities for division and concurrency of labor.

Change

If the profound changes were Newtonian, then we would have the information to forecast what the system was going to do for the future, and we could create optimum systems and optimum plans.

It is necessary to challenge the idea that if everyone in a system optimizes his or her own performance, the system will work best.[1] What we have to remember is that a system consists of repetitive tasks involving people, machines, materials, and methods, as well as the environment in which they work. The result of all the tasks is, or should be, movement of the system toward its aim; thus, each element has to be thinking about the whole. For example, the body consists of billions of independent cells which all work for a common purpose—yet clearly each one is not working only in its own self-interest.

Managers today are forced to optimize their own department or area at the expense of the system (pulling in sales to meet monthly quotas and creating boom/bust cycles; reducing inventories at cost of availability for customer; releasing products on the promised delivery date but with a list of known bugs). They look after their own career rather than doing what is best for the company overall—holding back information, using people, playing politics, creating fires so they can be put out, building empires, creating win-lose situations.

The holistic or systems point of view is the opposite of the reductionist notion that one can understand how a system works by completely understanding its component parts. Some of the vital information about a system always remains unknown and unknowable; therefore, it is not possible to completely understand the relationships among all the parts. Understanding a system in a holistic manner means

Notes for this chapter begin on page 147.

looking at the system as a whole; one can look at a forest, with its various fauna interconnected in a supporting environment—or at the human body, with its many different types of tissues all interacting with and dependent upon one another.

Many corporations today, driven by investors and analysts to optimize the bottom line in the near term, are typically laying off people, selling inventories at discounts, and are moving good managers to other areas needing help, replacing them with others who may not be as proficient and who have to deal with the challenges of adapting to different cultures and systems. Corporations are reducing quality by purchasing lower-cost components, and are antagonizing employees to force them out. This may improve financial results in a short time, and make a few highly visible "heroes" very rich. But the "cure" may cut too deeply and leave the company even less capable of providing long-term value and grade of product to its customers.

Despite lip service to the importance of the customer, observation still shows that most companies concentrate more on what they get from customers (money, profits) than on what they are going to provide to the customer (the quality of the product, and extensions to the product such as spare parts, courteous support, a product line that grows with the customer's needs, and so on). But Deming's profound changes tell us we should look at what is produced and how well we are providing the product or service, rather than selecting the products to offer the customer by the profit those products will make. An example of ignoring this would be if you ran a drug store and concentrated only on products with a high profit and stopped selling the products with a low profit—even though the customer needs a whole line of products from the drug store.

7.1 Excuses Commonly Given for Decline in U.S. Competitiveness

United States industry is criticized for its lack of success in the global competition of the eighties and nineties. This criticism comes from business journals, newspapers, television programs, and especially from the market analysts who serve Wall Street. Each constituency has its own set of complaints, concerns, and many offer a program of solution. One article explained the issues this way:

> *The critical task for management is to create an organization capable of infusing products with irresistible functionality or, better yet, creating products that customers need but have not yet even imagined.*

> *This is a deceptively difficult task. Ultimately it requires fundamental change in the management of major companies. It means, first of all, that top managements of Western companies must assume responsibility for competitive decline. Everyone knows about high interest rates, Japanese protectionism, outdated antitrust laws, obstreperous unions, and impatient investors. What is harder to see, or harder to acknowledge, is how little added momentum companies actually get from political or macroeconomic "relief." Both the theory and practice of Western management have created a drag on our forward motion. It is the principles of management that are in need of reform.[2]*

Top management are increasingly vocal in their responses. They are misunderstood, some claim, because their companies' problems were not caused by them; others accept responsibility and start providing leadership on the problem. Especially among the larger firms, however, blaming others is the more typical motif: here are some of the "culprits" and the generic maneuvers used to deal with them.

Labor
- Get rid of expensive older employees, and hire cheaper young ones who won't get pensions.[3]
- "Out-source" jobs and functions that until now have been done by your own employees.
- Replace employees lacking education rather than retraining them.

- Hold employees accountable for quality and productivity increases.
- Increase automation.

Domestic Competition

- Obtain exclusive government concessions, tax breaks, and grants, if not solely for your company, at least for your industry (e.g., petroleum).[4]
- Enter into "strategic alliances" with competitors, in the hopes of blunting or wresting away their advantages.
- Imitate "successes" of others, or perhaps just their rhetoric.
- Limit entry of new competitors.

Foreign Competition

- Restrict entry by treaty.
- Accuse of "dumping" when they sell below your price in "your" market; seek punitive duties to punish them (thereby punishing the American consumer).
- Position your own products with "patriotic" themes.[5]
- Blame foreigners for trade "deficits."[6]
- Restrict sales of foreign companies by tariffs, quotas, and domestic-content regulations.
- Continue to use offshore companies privately as suppliers while bad-mouthing them in public. (Chrysler is a prime example, but far from alone.)[7]

Customers

- Patronize with imitations of more desirable foreign products.
- Use government to tax those who buy foreign products; for example, use penalties on Japanese manufacturers who produce domestic cars here with U.S. parts and labor.
- Harass with emotional pleas to "Buy American" to save "our" jobs.[8]

At some point many people begin to see through the excuses that are implied in these actions. Some people vow, for example, not to buy an American car again until our producers stop making excuses for their unwillingness to produce more of the cars Americans want. But few people understand, in other than a personal way, why the excuses don't hold water. Our purpose here will be to try to hold several of the underlying excuses under a strong light and examine them in the context of the wisdom of Deming's profound changes.

Lack of Automation and Robotics

Almost everyone knows now that most of the world's industrial robots work for other than American industry. Most American management would like to have robots because they believe that:

1. Robots do away with labor problems because fewer employees are required, chiefly as custodians of the robots.
2. The foreign industries whipping U.S. companies are typically doing it with automation.
3. Robots are therefore a good solution to our problems.

Let's take a look at each of these arguments and see if they stand up:

1. Robots, if intelligently used, do save labor. But they can actually *increase* labor problems because the employees who deal with the robots must be highly trained and motivated, must know when and how to intervene in the robots' work, and must cooperate well in a (sometimes) smaller workforce. Successful application of robots and automation requires that management first follow Axiom 1 of Deming's philosophy which lists management's first job as gaining leadership and cooperation. This has not been done in most U.S. companies.
2. Books and articles written by Americans about industry in countries whose companies are beating ours in competition feature declarations of amazement at how *few* robots were in evidence in their plants. The same authors—accustomed to the disorder, complexity, and confusion which so often reign

in our factories—remark at the cleanliness, organization, and lack of drama in these companies' operations. Only recently have U.S. companies started to look to reducing complexity in their operations *before* bringing in automated tools. Overcoming the American penchant for gadgetry and addiction to crisis is a necessary early step on the long road to being competitive with automation.

Where robots and automation are used successfully, it will typically be found that management have attended to their responsibilities under Axioms 2 and 3, and have successfully used systems and division of labor for doing work. These are *prerequisites* to automating, not something to be thought about later.

3. Robots are not a solution in the sense that they will *cause* our problems to go away; they are an effect—an eventual *result* of management's isolating and eliminating complexity (Axioms 4 and 5) and establishing a secure environment. This type of environment is one in which people at all levels can safely contribute to the company by continually learning, as well as by transforming their subjective knowledge (know-how) into more objective forms that are available throughout the organization (Axiom 6). Unless this is done, automation is doomed either to failure or to minimal return.

An evolutionary learning process must be set in motion in order to continually remove complexity from operations. Prematurely installing robots will merely automate the complexity making it *more* difficult to remove.

Cultural Differences

Travel to other countries isn't necessary to understand that their cultures differ from ours. Cultures can differ greatly within a country, a state, a city, or even a single company. People do not have to speak another language or wear different clothing to be in another culture. You can walk fifty feet and move from one culture into another. Sales and manufacturing are two different cultures, as are marketing and finance.

In the early post-war years, when industry was in shambles, Japan produced at low cost and often at low quality. Many in the West ignored the possibility that her culture was capable of anything else. They would soon be shown to be in error:

> *For Japan, the era of competing on the basis of being the low-cost imitator of the West and of using exports to stimulate its domestic economy came to a dramatic end with the rise of the yen. Japanese companies reacted by shifting the competitive battleground to a different plane. They have moved up-market to compete on the basis of quality, innovation and product leadership. . . .[9]*

Most of the excuses based upon cultural differences are no longer offered by those who have actually been to Japan and gained some depth of understanding of how their industry works. A few characteristics emerge that make Japanese companies really different from ours:

> **learning,** spurred by recognition of the limitations of current knowledge
>
> **continuous improvement,** reduction of complexity and variation
>
> **vision,** the wisdom to choose appropriate goals for the long-term future, not merely for the next financial quarter or year
>
> **execution,** taking the vision seriously enough to reach it through their actions

Yet none of these is unique to either culture. In Japan, hierarchy is almost a genetic characteristic, so wherever one is in that hierarchy one has no need to secure one's territory; Americans deny their hierarchy and then proceed to consume themselves trying to establish it.

Japanese companies such as Honda have shown that these characteristics can be applied even within the United States, in organizations where only the top manager is Japanese.

It seems time to ask why our culture isn't an *advantage* to us. Our forefathers braved an uncharted ocean, fought off Indians, challenged the British crown which had established the stronghold in the New World, created a constitution which for the first time in history of man had as its express purpose the *limitation* of governmental power, provided for a sound monetary system (the gold standard), kept out of the way of the operation of the market, increased their territory many times over, and became the wonder of the world. Talk about vision and execution! Yet since the mid-nineteenth century the United States has slowly been dissolving. Today we are the largest debtor nation in the world; historians have lost track of our panics, recessions, and depressions; our top executives lament our so-called "negative" trade balance and try to get us to buy their products as some kind of quasi-patriotic act; we elect politicians whose personal scandals have become as commonplace as their disregard of the system's original aims; and most people would be hard-pressed to mention anyone whom they consider a statesman. Our culture used to be an advantage to us—why is it no longer?

Trade Barriers

Trade barriers have long been instruments of diplomacy and foreign policy in many countries, and they are very much in use today. When a nation erects barriers to international trade, its policies are said to be *protectionist*, yet such protection is always for the benefit of a privileged few—a classic case of suboptimization of a system. As we did with Deming's and Taylor's philosophies, let's examine where protectionism and its tools, trade barriers, come from.

A zero-sum paradigm from the seventeenth century called *mercantilism* has helped shape trade policy from the sixteenth century to the present day, even though few would recognize the term. Mercantilism is a theory based on the belief that the total volume of trade in the world is fixed—a zero-sum game requiring government intervention to "safeguard" domestic producers against foreign competition. Trade surpluses and the resulting net inflow of precious metals are sought through tariffs and other barriers to discourage imports, and subsidies to encourage exports.[10] Economically bizarre, mercantilism must be viewed instead as a political theory in support of imperial State power, as well as a special subsidy and monopolistic privilege for favored domestic producers. Because subsidy and privilege can only be conferred by the government at the expense of the remainder of its citizens, the majority of consumers have been losers under mercantilism.

Wars have begun over restrictions of trade between nations. Foreign exchange offers a source of revenue from duties imposed on it, as well as a means for government advancing its control over its citizens, and a medium for granting favor and privilege. It can be an invaluable source of advantage for a company or industry to have its market protected by the State. Consider that the British Corn Laws, by regulating the trade in grain (including wheat), protected British growers at the expense of higher bread and cereal prices for the British worker from 1436 to 1846—a period of over four centuries.[11] Modern examples abound.

Here are further issues to ponder and discuss:

- Why are *our* trade barriers fair and reasonable while theirs are unfair and inappropriate?[12]
- Only a few years ago the nation with the most restrictive trade barriers was France. Now this dubious distinction belongs to the United States.[13] Harley-Davidson, for example, now touted as a quality "success story," survived what should have been its own demise in 1982 by obtaining from the U.S. International Trade Commission a huge five-year increase in tariff duties on all Japanese motorcycles of 700 cc and over engine displacement. At the end of the five-year period Harley-Davidson requested that the tariff be removed.[14]
- Many nations hostile to ours, such as Communist China,

have been given "most-favored nation" trading status by our government.[15]

- Why does American industry consider any particular market or segment so vital to its success that effective competition will not be tolerated if it can be put aside by political expedients?
- How shall we account for the foreign companies that sell well in other countries despite the trade barriers? Are they getting special treatment not accorded to U.S. companies?
- Why can't we *expand* the market instead?

One scenario that makes visions of trade barriers dance in the heads of Western business is that of foreign producers becoming more efficient across the board in every product "we" make—with advantages in labor skills, accumulated capital, and natural endowments. Couldn't this ultimately mean that America, the emancipated colony, would revert to its former status, supplying raw materials to other countries and importing them back as finished goods?

The answer depends on management. Ideas have consequences, and the ideas that comprise the managerial paradigm will determine how management would react to such a condition. The typical reaction to foreign superiority has been to erect further trade barriers; let us see if there is a better way.

Assume for a moment that the United States is flanked by producers who are more efficient in every regard. The British economist David Ricardo (1772–1823) grappled with just such a problem in international trade, but addressed it as an issue in division of labor.[16] Assuming that goods, but not workers or the factors of production, could freely pass between nations, Ricardo demonstrated that a win–win solution to this problem is possible, and can best be achieved if each producer concentrates on those commodities *for which its superiority (relative advantage) is the greatest*, leaving to other producers those commodities in which its superiority is less. This effect is called the law of comparative cost. It, of course, doesn't guarantee that every nation will be rich; as the natural distribution of land, labor, and capital is inherently unequal, no method can seriously make such claims.

For some, Ricardo's law is counter-intuitive in that it would mean leaving more favorable domestic facilities unused in order to buy from areas where conditions are comparatively less favorable, concentrating on those commodities in which one's superiority was the greatest.[17] In addition, it also involves cooperation and the removal of barriers to trade.

Another paradigm, that of the "balance of trade," a carryover from the days of classical mercantilism, is still quite strong in most of the world. Let's consider that trade consists of the sum of goods, services, and money exchanged between one nation and another. Now, logically, if the sums have been done accurately, the items given and the items received are always exactly equal—in balance. Trade balances are always equal.

But the legacy of mercantilism, anachronistic at best in this century of unbacked paper currencies, is to demand a net inflow of monetary receipts over monetary payments; this is called a favorable balance, or *trade surplus*. If monetary outflow exceeds inflow, this is called an unfavorable balance or *trade deficit*.

What this ignores, of course, is that at the time of each individual transaction between the two nations, each party considered the transaction more to its advantage; one preferred the goods, money, or service it received more than whatever it was giving up. Counting only the monetary items exchanged serves no purpose; no one ever pays out money unless one prefers what one gets in return.

"Deficit" and "surplus" thinking in matters of trade, then, are not only anachronistic but a delusion:

> *Most people believe that U.S. trade deficits are economically unhealthy, and that the Japanese are to blame. Yet our trade deficits actually are an unearned blessing for which we should be thankful. Trade deficits bring economic growth and health.*
>
> *Capital investment is the fuel of that dynamic engine called capitalism. Accumulation of capital and its efficient use,*

which are inherent aspects of free market economies, are to a great extent responsible for the high standard of living the masses achieve in a free economy as compared to the Third World or socialist economies. But how is capital accumulated? By savings that are then invested in productive facilities.

Our spendthrift habits and Federal deficits drastically reduce America's ability to accumulate capital. But the Japanese (and others) have come to our rescue. Where we have neglected our own economic health, they have invested in it. In fact, the Japanese are using their hard-earned savings to bail us out of our own economic folly. To a significant extent, they are fueling the capitalistic engine that maintains our lifestyle.

Net foreign investment flowing into America and our trade deficit are the two sides of the same coin—one necessitates the other.[18]

Barriers to trade are suboptimizing because they obstruct not only the international operation of the principle of division of labor as applied in Ricardo's law of comparative cost, but also today's international movement of capital (foreign investment), fairly rare in Ricardo's time.[19] Economist Milton Friedman implicitly observed this suboptimizing nature when he said,

The trade deficit is a sign of American strength because its counterpart is a capital inflow. The Japanese and other foreigners, as well as domestic residents, are investing in the United States because they can get a higher return here than they can elsewhere. Is that a sign of our weakness and their strength or is it the other way around?[20]

Cooperation, as Deming says, is an alternative to considering others as rivals in a struggle for limited resources. The law of comparative cost shows how the principle of division of labor, already described by Adam Smith, is valid even across national borders so long as those selfish producers anxious to justify "protection" have not prevailed in the councils of State. Counter to "balance of trade" thinking, the gains from the division of labor are always mutual.

Government Interference

1. **Anti-Trust Laws.** We would be unrealistic to ignore the way a century of this government interference has systematically put American companies, and their customers, at a disadvantage. Yet in current economic conditions the Justice Department is less likely to make trouble for companies that collaborate on R&D and other activities which improve competitiveness instead of merely trying to limit markets.

2. **Regulation.** Few American industries have escaped burdensome regulation which increases the cost and complexity of their products and often adds product features not called for by the market. Other countries are increasing their regulatory activity, however; this may equalize to a great extent the competitive differences due to regulation.

3. **Using the Government.** For centuries business has sought the protection of the State—from violence, from unfavorable laws, and from domestic or foreign competition. Using the power of government to interfere with or eliminate competition—by granting exclusive franchises, growth incentives, subsidies, bailouts, monopolies,[21] tax advantages and incentives, outright funding, and "saving" selected industries and jobs—has often been preferred over earning one's place in the market by producing things buyers like.

The typical person on the street knows little of this special treatment. A close look at the ways by which most governments, including that of the United States, have favored certain industries, cartels, and individual interests at the expense of the ordinary citizen or businessperson, would shock most people.[22]

W. Edwards Deming assured Japanese industrialists in 1950 that, if they followed his teaching, in a few years other countries would be crying for protection from their goods. Note that he didn't say the *citi-*

zens of the other countries would be doing this—their governments and industries would. The citizens would be enjoying the products and asking for more. Whom, then, would their governments be serving by this "protection"? They serve a small subset of the country's businesses who cry for suboptimization of the national system for their private benefit. We must remember that any time someone is "getting something for nothing" from the government, someone else is getting nothing for something.

In spite of their extensive (and expensive) lobbying in Washington through political action committees, American companies have not succeeded in banishing serious foreign competition. Why? One reason might be that the lawmakers have a little more respect for the citizens than does the business—and the citizens increasingly do not want tariffs and quotas to stand between them and what they want to buy.

Lack of Employee Motivation

What is motivation? When management use the term they often mean "seeking the interests of the company," or more cynically, "seeking *our* interests." In the former sense, when have employees *ever* been motivated? In the history of the world "motivated" employees have probably been notable exceptions to the rule, and in each case something surely was better in their environments when compared to the norm.[23]

Deming, of course, relates this directly to the actions of management, who are responsible for the system in which the employees work. Management have probably had exactly as many motivated employees as they deserved. This book contains a veritable catalogue of modern management actions that destroy loyalty and motivation.

Given the unmotivating systems in which most employees have worked, their accomplishments are amazing. Even more amazing is the degree to which these people, "unmotivated" by day, sit on school boards and other community organizations at night, in some cases in addition to functioning as single parents. Clearly these people *are* motivated.[24] The chain of extensions that could have been achieved with Deming-based systems challenges the imagination.

As if their performance were not troubling enough, clear signs are visible today that managers are adopting an even more appalling scheme for motivating employees: scientifically applied fear. Thanks to the prescriptions of a few psychologists hired by management, employees have been declared guilty of the crime of "entitlement": they "believe they do not have to earn what they get" and are owed their salaries "because of *who* they are, not because of what they *do*."[25] Management must break them of this habit by instilling and carefully tuning an atmosphere of fear and anxiety throughout the company. Because of its suboptimizing effects, its base view of people, and its anti-learning bias, this bizarre logic violates virtually every element of Deming's system of profound knowledge.

High Cost of Capital

Being artificial creatures, corporations require capital to come into existence, and those supplying the capital (by purchasing stock or lending money) do so in anticipation of a return on it. Although debt is usually preferred over equity because of the tax advantages it confers, obtaining capital by either method is risky because the investors or their agents will usually try to influence the actions of management to their own advantage—another case of suboptimization.

Borrowing capital—now not only endemic to American business practice but insinuated into American culture as well—is subject to much the same hazards as equity transactions. Although the lender has no *legal* rights of control, many loans, especially large ones, have provisions for being called in if the lender is not pleased with management's actions.[26] In addition, the borrower wants to maintain a credit rating for future loans and may yield to the lender's suboptimizing influence even if not in any trouble.

Henry Ford, known for his hatred of bankers and others who could gain control of his company, declared, "Any business is finished when it begins to finance." Although Ford had many faults as a CEO, his avoidance of outside control of his company provides a paradigm all companies should consider. He relied instead on his assembly lines to provide capital in any amount needed. Keeping your company out of the control of outsiders by using its productive operations as your bank is an optimizing strategy because it encourages innovation and productivity improvements by digging into the gold mine of opportunities that most companies contain within themselves. In addition, solving productivity problems is cheaper and much less risky than large-scale capital investment.

New capital may not always be needed. Through innovative thinking and reducing complexity, using existing facilities may be practicable instead of buying new ones. Existing capital is not always well maintained or exploited; expensive equipment often sits idle while waiting for input from upstream production bottlenecks. Here again, Deming's profound changes provide a way to gain the knowledge needed to improve.

Employee Education

As an excuse for decline in their competitiveness, companies have cited low levels of employee education and skills, in spite of billions spent on corporate education and training each year. Why? Here are some reasons:

- Education has often not been considered an investment.
- Employers have neglected broad, basic education, such as in philosophy and how to think, to express one's ideas, to speak another language, to understand another culture, to experiment. Company education has concentrated instead on skills-training.
- The education function is often considered a temporary assignment for professionals who do not know how to teach.

- Companies are often unwilling to spend the money to bring in the right teachers if unable to teach a needed course in-house.
- Companies lack desire to form education partnerships among themselves.
- In many cases corporate education has become yet another bloated bureaucracy, more concerned with its own care and feeding than with the constituencies it was established to serve.

But these may all be moot points. With the advent of what may well be the corporate form of the future—the holding company—just the opposite is happening: workers increasingly are expected to have the required education or skill as a condition of employment (or contract). Corporate education is becoming more propagandistic—a channel for announcing policy, promoting new values, and aiding other manipulations of the work force. Contrast this with the thought of the economist Alfred Marshall:

The most valuable of all capital is that invested in human beings.[27]

How do these several excuses for decline hold up? Not well, in our opinion. One by one they wither in the light of critical enquiry. Management have little to hide behind. For the good of their employees, their customers, and for the country as a whole, management have *no excuse* not to learn and apply Deming's profound changes.

7.2 Understanding Change

Throughout this work is the underlying message that management, coupled with entire organizational systems, must change. Yet change is hard. In 1513, Machiavelli wrote in *The Prince:*

It must be remembered that there is nothing more difficult to plan, more doubtful of success, nor more dangerous to manage than the creation of a new system. For the initiator has the enmity of all who would profit by the preservation of the old institution and merely lukewarm defenders in those who would gain by the new.

Of the changes going on in our world, three types can be identified. *Evolutionary* change and *revolutionary* change are the two most commonly thought of. We will focus on the third type, though, and that is *change in the process of evolution*—or said another way, change in the process of change itself. This third type of change has been studied and recommended for action by Walter Shewhart and later by Dr. Deming.

Because of the orientation toward the first two types of change, we hear a variety of excuses for why there has been a decline in U.S. competitiveness—cultural differences, trade barriers, government interference, high cost of capital, lack of employee motivation or education . . . and more. The litany is recanted from Fortune 100 CEOs, to the President's Commission on Industrial Competitiveness, to MIT's Commission of Industrial Productivity. They are only excuses, pointing the finger at something or someone else. None of these reasons gets close to the system of causes of the decline. Too few leaders of organizations today have reached the point of transition beyond classical problem solving to consider the power of change when it is change in the process of evolution itself.

To understand the third type of change, it is useful to identify the phases of learning we go through as we become more proficient at comprehending and handling problems.[28] In the earliest phase we learn to simply *solve* problems by rote. We look at the model, formula, or structure and apply straightforward mechanical principles to gain the solution. In the next phase of learning, we come to the point of being able to *define* problems. We gain the experience and knowledge needed to pick out problems from non-problems based on comparison of results between what we expect versus the actual. We develop hypotheses about how processes work, and analyze the process results to test our theories.

A further phase of learning is one of *questioning* problems—where we seek to understand the source of the problem and what led the problem to come about. Here we may use an approach of seeking out contradictions between the theoretical model and actual operation of a system, then strive to understand in what ways the contradictions come about so that we can more fully understand the system itself. At this stage, the scientific method is applied. We reach a level at which we can understand the process itself and add back into it what new knowledge we gain (or even completely change the process when we find it no longer faithfully represents our observations of reality).

Ultimately, one reaches a state of adopting a principle of learning.

The point of looking at learning as having phases is to start to appreciate how *change* requires gestation. Time will pass, and growth of knowledge will occur at individual and organizational levels before change truly happens. While one can argue change is always occurring, it is not automatic—nor is it always for the better, given variation and randomness in systems. For those attempting to promote change, understanding the various elements of change will help them bring about change (and improvement) more swiftly.

Change can be looked at as a hierarchy, with each level of the hierarchy increasingly more complex relative to how a particular change at that level gains acceptance. The following goes into the details of each level of the hierarchy of change.

Changing a physical thing. This is a level we are very familiar with, and change at this level occurs daily. We change the clothes we wear; we eat a variety of foods; the news broadcast is different each day; the weather changes. In fact, in many instances we experience change in physical things without even noticing a change is taking place. When we do notice, in most cases there is little, if any, hesitation about the change.

Changing the *way* we do things. We are also constantly changing the *way* we do things. We discover new tools and processes to use, and find that using them makes a difference in the result we are looking for, be it a new writing instrument, a new computer application, a new route to work, a change of hair style. When the result is viewed as beneficial, we implement the change; otherwise, we revert to what we previously did. Again, many times a change in the way we do something may not even be something we take great note of at the time.

Changing our mind-set and theories. Here we are at the level of thinking about what we do—this is the metaphysical level. We think about why we choose a particular process; we think about our motivations for a variety of behaviors and choices we make. We take into consideration many factors, tangible and intangible. From this we ask ourselves if our actions are consistent with all we understand and believe. Having done our analysis, we choose either to change our perspectives and theories or to continue as we were. It is important to see we are referring to a model of active change here. Note how important changes in mind-set are actions done consciously, rather than just acceptance of any new ideas that might come along. At this level, while the analysis may be frequent for some, actual change usually is not— only happening on the order of a few times a year. It is still fairly easy for this level of change to occur, though we are more conscious of it occurring. But the opportunity for change comes less often as more factors are in operation.

Changing our relationship with the universe. This has to do with *how* we learn about the nature of things—epistemology. Our learning leads to greater awareness of the universe (probably best characterized in terms of "society" or "the world") around us and our interaction with it. We tend to see a change at this level occur only every decade or so, as the change occurs across a large enough group of people. Examples which indicate these relationship types of changes include the evolution from corresponding with friends by letter early this century to use of the telephone; the bulk of public news distribution evolving from newspapers to television; extended travel by train, boat, or car shifting more and more to jet airliners; and a greater awareness developing of ecological systems. The acquisition of enough new knowledge (coupled with the development of technology resulting from that knowledge) forms the tools—both physical and intellectual—that bring about these changes. With these changes more knowledge is gained, bringing about new tools which drive further changes and growth in knowledge.

At this level, Shewhart, Deming, and others provide us with the scientific method for acquiring knowledge. Most often thought of as being used in our formal work—but applicable to learning in all facets of life—this method provides us with an improved means for understanding what goes on around us. Starting with theories, we observe, understand, improve upon the theories and repeat the process—a cycle that must continue without end. The scientific method provides a systematic approach which accelerates the learning process inherent in understanding and improving (changing) our relationship with all that exists around us.

Changing what we think the universe is. How we see the universe—our cosmology—only changes over a very long period, and is met by much resistance until there is sufficient, compelling evidence. Copernicus challenged the notion of the earth being at the center of the universe and was excommunicated by the Church of Rome. Galileo's refinement of the telescope (also challenged by Rome) provided people with the information to verify Copernicus' observations. The acceptance came about only over many scores of years.

Physicists such as Werner Heisenberg, Erwin Schrödinger, Percy Bridgman, and Albert Einstein took the world's perspective from a Newtonian view of events to that of quantum physics and provided insights that led to significant changes in how we see the universe. They provided us with the *fundamental* insight of the nondeterministic universe, not totally predictable using current levels of knowledge. We have come to understand that variability (randomness) exists within, and affects, all systems.[29] Therefore, we will always benefit from efforts to better understand the causes of that variation for the sake of improving the affected systems. With that in mind, Deming also teaches that we and the universe are not separate systems. His is an open, broad, and all-encompassing—versus closed and locally optimized—perspective in dealing with systems.

Changing our values. Reaching this level, it becomes clear that change occurs *very* slowly—the least frequently of any of the above. For instance, over two hundred years ago Adam Smith observed how self-interest drives a society's economy.[30] That value is still a key driver of economies even today. (It is interesting how Smith's observation can be viewed narrowly and negatively for the individual, becoming an excuse for "local" behaviors which suboptimize systems, or broadly and positively for its impact on society overall.)

A variety of elements come into play relative to understanding our values and how they affect our behaviors. We explore the values we have toward people in general and ourselves in particular—our motivations and our interdependencies. We establish a perspective on how much we consider ourselves to be "in the center" of the universe surrounding us—how we influence it, and are influenced by it. In the workplace, people must consider how they view the manager/worker relationship, suppliers, customers, competition versus collaboration (both internal and external to the organization), markets—as well as the more traditional processes and systems. Deming challenges us to view all of these elements, and more, in a holistic way, to value each part for the integral and unavoidable impact it has on the eventual outcome of any undertaking.

With all this, one starts to get a sense of the dimensions of change—to see where a particular desired (or inevitable) change falls in a spectrum and to appreciate how readily it will be accepted and accomplished. Understand that change is not something that occurs in only a passive way. With focused effort, people can bring about change themselves and accelerate what would otherwise be natural evolution.

Again, beneficial change comes as a result of learning, from gaining knowledge. This can be very clearly interpreted from Deming's teachings. Once new knowledge is gained, an individual/group/organization is forever changed. Significant enough change results in a "paradigm shift." What we try to bring out in this book is that learning and change go on at many different levels—some very local, some all the way to virtually encompassing everything. Having such a broad perspective helps lead the individual to a certain understanding about the complexity of systems and what all must happen to bring about long-lasting system improvement.

One finds making the most important and fundamental changes for business and organizational systems improvement will take a great deal of time and effort. There will be many forces along the way directly and indirectly opposing change. Fortitude and faith become the ultimate ingredients of seeing change through! Change is not easy, and there are few who will exhibit the courage to make a difference. Akio Morita, the president of Sony, said in 1987, "I am afraid that American industry has lost faith in itself and that the trade imbalance will not be corrected until that faith is regained." Will he be right?

There is reason to have confidence, though. By understanding the nature of change and helping others understand the change going on around them, leaders can guide their organizations toward regaining the faith necessary to encourage further change—not only for the sake of improving productivity and rebuilding industrial strength, but for regaining and securing our high standards of economic success and national presence within a cooperative, global community.

NOTES

1 Adam Smith, usually referred to as the source for this idea, was describing the essentially unorganized behavior of a free marketplace, and how the self-regulating market requires producers to act in other peoples' interests in order to stay in business (the so-called "invisible hand of the market"). Especially for organizational systems, we believe Deming argues for even more-explicit cooperation than is usually attributed to Smith's model.

2 C. K. Prahalad and Gary Hamel, "The Core Competence of the Corporation," *Harvard Business Review* (May–June 1990), pp. 79–91.

3 See the description of portfolio careers in section 2.3, "Taylorism and Neo-Taylorism."

4 Even regulation, which big business has traditionally opposed, may sometimes be preferred today because it gives scope to the desire to limit entry of new competitors, and enjoy government-assigned markets instead of having to earn them by system-optimizing activity. Deming's counsel is to collaborate, expand the market, and then compete. Unhappily, such collaboration is often forbidden by law.

5 Chevrolet and Wal-Mart are two examples. See Steve Lohr, "New Appeals to Pocketbook Patriots," *New York Times*, 23 January 1993, p. 37. The article also contains a rebuttal by Louis Stern, professor of management at Northwestern.

6 Humorist Dave Barry lent keen insight on President Bush's trip to Japan with the Detroit auto executives "who had to be flown over in huge military cargo jets because ordinary planes would have been unable to lift their wallets. The goal of this trip was to get Japan to import more American-made cars, but the U.S. delegation sounded like a bunch of big fat overpaid whiners. A lot of people thought it was pretty embarrassing, seeing as how the biggest problem facing the U.S. auto industry is that a lot of *Americans* won't buy American cars." (*Dave Barry Does Japan*, Random House, 1992; pp. 102–103.)

7 For several years General Motors has been the number one importer of vehicles into the United States; Ford and Chrysler are number three and number four. In 1992 Honda built more cars in the U.S. for sale here than Chrysler did. (Walter Huizenga, "Stop Dumping on Detroit . . . But It's Living in the Past," *New York Times*, 14 February 1993, p. 11.)

8 Such propaganda is effective: in November 1992 Yankelovich Partners, a market research firm, found in its annual survey that an amazing 61% of people surveyed "feel guilty when purchasing non-American products." (Steve Lohr, "New Appeals to Pocketbook Patriots," *New York Times*, 23 January 1993, p. 37.)

9 Tomasz Mroczkowski and Masao Hanaoka, "Continuity and Change in Japanese Management," *California Management Review* (Winter 1989), p. 39.

10 Some background is provided in "Strategic Trade Policy and Mercantilist Trade Rivalries," by Douglas A. Irwin; *Journal of the American Economic Association*, Vol. 82, No. 2, May 1992. Today's governments have all thrown off the discipline of precious metal money; those still practicing restriction of

imports, of which Deming says the United States is the most restrictive of all, should more accurately be called neo-mercantilist. Today the multinational firm has a vested interest in a decline of mercantilism. See "Whether Free or Fair Trade, Corporate Mercantilism Rules the Day," by Peter W. B. Phillips, *Challenge* (January/February 1992), pp. 57–59.

11 In 1838, a group of merchants and manufacturers founded the Anti-Corn Law League in Manchester to protest this protectionism. Over the next fifty years this group (the so-called Manchester School), led by Richard Cobden (1804–1865) and John Bright (1811–1889) succeeded in moving Britain from 350 years of mercantilism to a more *laissez faire* system. By the turn of the century, however, the gains from this philosophical revolution had largely been reversed by the Conservative Party and the Fabian Socialists.

12 For a revealing account of what U.S. regulatory laws cost our own companies—ultimately to our own handicap in world trade—see the article "America's Real Trade Enemy Is Washington," *Wall Street Journal*, 2 March 1992. Also, in a 1990 *Report on Unfair Policies by Major Trading Partners*, the Japanese Ministry of International Trade and Industry found the trade policies of the United States to be the most unfair.

13 Cited by Deming in *The Deming Library Series*.

14 Phillip Oppenheim, *Japan Without Blinders: Coming to Terms with Japan's Economic Success*, Kodansha International, 1992, p. 83.

15 See "We Win No Oscars for Tibet," *New York Times*, 13 April 1993, p. 21. The title refers to the U.S. betrayal of a friendly nation, Tibet, into China's brutal domination as a way of "sweetening" the trade deals with China.

16 Ricardo's law of comparative cost, or comparative advantage, will be discussed further in Chapter 8, on the Axioms (Axiom 3, Division of Labor). See his book *The Principles of Political Economy and Taxation*, published in 1817. Since Ricardo's law is so central to social cooperation it is also called the law of association.

17 For a more complete explanation and a worked-out example, see Ludwig von Mises, *Human Action—A Treatise on Economics* (3rd rev. ed.), Henry Regnery, 1966, pp. 157–164.

18 John D. Fargo, "Thank the Japanese for Our Trade Deficits," *The Freeman* (October 1991), pp. 388–393.

19 In general, the material standard of living of a country is based on the amount of capital invested per capita. In addition to making it more difficult or more expensive for consumers to obtain foreign goods, protectionism also stifles foreign investment of capital which could increase the standard of

living in a country. Instead, foreign investors are sometimes treated as exploiters who ought to be expelled.

20 "The Adam Smith Address: The Suicidal Impulse of the Business Community," *Business Economics*, January 1990, pp. 5–9.

21 Monopoly is a technical term of economics meaning a market in which there is one seller. As long as there is freedom of entry, a monopolistic situation can be terminated by the entry of another seller. Government-sustained monopolies make it illegal for new sellers, who might serve consumers better, to enter the market.

22 Antitrust regulation and other governmental interference with market processes are discussed in "Regulation in America: Is This Still the Land of the Free?" *The New American*, special issue, 17 May 93.

23 Section 2.2, "Japan's Important Change," describes what is probably the outstanding case in history.

24 The issue of motivation is taken up again in section 11.3, "The Restoration of the Individual."

25 Judith Bardwick, *Danger in the Comfort Zone—From Boardroom to Mailroom: How to Break the Entitlement Habit That's Killing American Business*, AMACOM, 1991. An especially disturbing book because it provides management who wish to do so with a pseudo-scientific pretext for using fear as a tool against their fellow members of the organization.

26 Because Mazda Motors (Toyo Kogyo Ltd.) had spent so much money developing the Wankel rotary engine in the early 1970s, only to be faced with the oil "crisis" of 1973, their creditor Sumitomo Bank stepped in and placed its own finance people as overseers of Mazda until the company became profitable again (2 1/2 years), reducing the rotary engine project to a tiny fraction of its previous size. This allowed a 26% share of Mazda stock to be bought by a foreign investor, Ford Motor Company.

27 From his *Principles of Economics* (8th edition), 1949.

28 Section 4.2, "Phases of Learning," parallels this discussion, with additional detail.

29 To complicate things further, even a deterministic, but nonlinear, system may behave in an apparently random manner, frustrating our attempts to learn about it.

30 Smith's own words from his *The Wealth of Nations* are still eloquent today: "By pursuing his own interest [one] frequently promotes that of the society more effectively than when he really intends to promote it. I have never known much good done by those who affected to trade for the public good."

Six Axioms for Comparing Taylor and Deming

8.1 Nature and Origin of the Axioms

Perry Gluckman describes the origin of the axioms this way.

Why were the axioms written? An axiom is a statement of something which is regarded as a self-evident truth. Axioms stick in people's minds. Most people can remember their axioms from geometry. For better or for worse, axioms are a good way of stating basic concepts clearly. The axioms for Deming's principles were written when we started to work in the product development area for Dow. When you work in the product development area, things work so slowly you have to have reference statements you can go back to in order to keep a check on what you are doing. Historically, one of the most effective ways is establishing axioms.

When Dr. Deming was reviewing Everyday Heroes, *he suggested it would be nice if we had a concise definition that showed the difference between his philosophy and Taylor's philosophy. At the time, most people were not aware of who Taylor was, and few people had an idea of what Deming's philosophy was. Instead of creating a definition that showed the comparison, I decided to develop a set of axioms that accurately described Taylor's philosophy in such a way that someone who practiced Taylor's philosophy would endorse it. With the help of Don Wheeler and George Watson, we now have sets of axioms for Taylor's and Deming's philosophies.*

One of the reasons the two sets of axioms are useful is because many of the practices Deming teaches can be Taylorized. It is important that the practices of the Deming school (control charts, JIT, quality circles) be used in ways that are consistent with Deming's axioms; otherwise, they will fall into Taylorism, and they will not yield the same results.

151

Let us assume, for example, we have adopted the use of control charts in the manner Deming teaches, for learning about our processes and classifying causes of variation (refer to Axiom 5). At some point management requests a presentation of our work. In the course of showing the charts, we are asked what *goal* we are shooting for, when the *results* will be obtained, and how much they will *cost*. Our answers to the effect that there is no end to reducing variation, and our assurances that it is always cost-effective to do so are not very satisfying. Management could very well give us two more weeks to finish the "project," at the end of which time they intend to reimpose their own specifications on the process, and require the workers to maintain control charts to demonstrate achievement of this goal. At this point our attempt to shift to a Deming paradigm has been halted with a return to Taylorism.

A group of workers, with management support, begins meeting together as a quality-improvement team. After deciding what they are going to work on, they decide that the process requiring improvement is in another area, one not represented on the team. They concoct a series of process changes along with a compelling presentation as to the importance of making those changes. The proposal is approved by management without hearing from the affected area, and the changes are made. The team is duly recognized for its "contributions" and disbands. Only later is it discovered that the area where the changes were made has had to increase its staff to maintain previous levels of production or quality. What started as a Deming-type effort to improve the process by gathering input at the worker level, supported by management, has turned into an example of Taylor's use of outside "experts" to create "optimum" systems.

As if that were not problem enough, let us suppose that this latter group becomes preoccupied with keeping statistics and records on the number of meetings it has held; it establishes a special area where bulletin boards are maintained showing the statistics; it adopts a catchy name, logo, and slogan; it occupies itself primarily with presentations to management and actually gets in the way of process improvement by gobbling up all the resources allocated to it. In addition to its earlier Tayloristic suboptimization, the group can now add extension transference to the list of problems it has created.

Many business and government leaders think they are already operating under Deming's philosophy; most are sadly mistaken.

It is important to reemphasize that Deming is partially an extension of some of Taylor's ideas. People *must* be aware of this lest they inadvertently Taylorize Deming.

An axiom is like a compass: it constantly tells you the direction in which to head. Deming's Fourteen Points and his Profound Knowledge are descriptions of his philosophy; but we found it was effective in applying his philosophy to have concise definitions that would provide constant guidance, like a compass, over a long period and through many different situations. The axioms are one way to provide this, and are valuable beyond the static exercise of comparing the philosophies of Taylor and Deming. They apply not only to Taylor's original system but all the more to today's entrenched philosophy of neo-Taylorism.

The entire set of axioms is presented here. Each is then discussed in its turn in the following section.

AXIOM 1: CONTROL OF A BUSINESS

TAYLOR

Control of a business is established by staffing positions of responsibility and authority with professional managers trained in the theory of scientific management and systems analysis.

DEMING

Control of a business is established by leadership and cooperation.

AXIOM 2: DIVISION AND CONCURRENCY OF WORK

TAYLOR

Improvements are due to management's increasing the division of work, and increasing concurrency (different aspects of work being done at the same time), within a project or among projects or processes.

DEMING

Improvements are due to increasing division of work, information, and creativity, and to increasing concurrency, within a project or process, or among projects or processes.

AXIOM 3: USING SYSTEMS

TAYLOR

Develop systems to perform repetitive tasks.

DEMING

Develop systems to perform repetitive tasks.

AXIOM 4: OPTIMUM SYSTEMS

TAYLOR

The optimum system can be created by proper formulation of the objectives of the system and evaluation of alternatives to meet those objectives. The information will be available to create an optimum system.

DEMING

No system is ever truly an optimum system: every system must be analyzed to understand the natural behavior of the system and the variation within it. Information for creating an optimal system is unknown and unknowable.

Axiom 5: Finding Causes

TAYLOR

Once a system has been properly defined and installed, any failure to meet stated objectives must come from outside the system.

DEMING

Inconsistencies and contradictions which become apparent upon analysis of the system may be used to detect and isolate the built-in flaws of the system.

Axiom 6: Role of Management

TAYLOR

Continuously monitor the status of the system for deviations from the system objectives to see if improper worker selection, poor motivation, inadequate training, or weak supervision are the causes of missed objectives.

DEMING

Create a secure environment so everyone can apply the first five axioms without fear. Offer support, reassurance, and appreciation.

8.2 Discussion of the Axioms

Axiom 1: Control of a Business

Taylor

Control of a business is established by staffing positions of responsibility and authority with professional managers trained in the theory of scientific management and systems analysis.

Deming

Control of a business is established by leadership and cooperation.

Axiom 1 is concerned with laying the groundwork for running any organization or business. Even on the face of it, Taylor and Deming are at odds:

	TAYLOR	**DEMING**
CONTROL	• Control is the *goal* • Management are the most important employees	• Control is the *effect* • Everyone is important
LEADERSHIP	• Goal = maximum efficiency of the defined system • Method = prescription	• Goals = • learn • help people • set the goals for the organization • Method = example
COOPERATION	• A *goal* to be enforced by standards and threats	• An *effect* of leadership which makes people secure

Control

For Taylor, management's first job was to be in control of the workplace. He believed in having a professional management staff, trained in scientific management, in all positions of responsibility or authority. Because of the great value he saw in this, Taylor set the stage for management salaries to rise disproportionately to those of the workers, even those workers who dramatically increased their wages by adhering to his "one best way" schemes. According to an October 1991 *Business Week* article, the typical ratio of salaries between CEO and factory worker in the U.S. is about 120:1, in Europe about 35:1, and in Japan about 20:1. Another article in the 18 June 1990 issue of *Industry Week* puts the U.S. ratio between 35:1 and 1,000:1. A 1991 *Harvard*

Business Review article noted that CEO Iacocca of Chrysler was being paid 34 times what his opposite number at Honda was receiving.[1]

In addition, Taylor used logical extensions by introducing a novel division of labor among management, creating eight new positions called *functional foremen*, who were required to carry out his efficiency schemes. This increased the complexity in both management and production, and was later extended further by Alfred Sloan at GM who greatly increased both the scope and power of the administrative and accounting functions.

Deming does not concentrate on control at all. Control is an effect of good management, and is not to be sought directly, but instead gained through leadership and cooperation. This is indeed a profound change to the current management paradigm.

Leadership

Taylor superimposed upon all management the goal of achieving "optimal" efficiency, and made it clear that getting to the "one best way of doing a thing" was not to be done in a series of small steps. Tayloristic goal-setting is rampant today, with managers often being expected to achieve the maximum results in a predictable, repeatable manner on the scheduled date. Viewing the worker as just another one of the machines in the plant made it easy for Taylor to adopt a leadership style of simply prescribing what is to be done and pushing workers toward it by means of threats and rewards.

Deming does not impose any set of goals upon management. Instead, he says that the first job of management is to set the visions and goals for the organization. The true leader will already subscribe to basic values and notions such as joy in work and treating people as full partners and must use the profound changes to introduce a habit of constantly gaining new knowledge in the organization. In both his latest books (1986, 1993) Deming cites Dr. Morris Hansen, his boss at the Bureau of the Census, as an example of a true leader. Hansen, who

Notes for this chapter are on page 171.

was just thirty years of age, gained an understanding of variation and statistical studies, understood the aim of his system, and realized that he could not achieve it by himself. He convinced experts to help him, and grew in knowledge and stature himself.

Cooperation

In the matter of cooperation the distinctions between Taylor and Deming are subtle: Taylor used cooperation as a lever to institute scientific management; Deming sees cooperation as a bloom of health in an organization, which occurs when management removes the barriers to it which they have erected.

Although for Taylor cooperation was a goal strictly instrumental to achieving the maximum possible efficiency, it was one of his four principles of scientific management and was to be enforced by standards, wage schemes, and even threats of dismissal. Cooperation was not to arise spontaneously among the workers but was to be planned and demanded according to the design of scientific management. Whereas we describe cooperation as the result of understanding a situation and the need for change before making a decision, Taylor simply mandated cooperation because he perceived it was a necessary condition for maximizing the division and efficiency of labor.

In Deming's philosophy cooperation is the natural and inevitable result in the company when management removes the necessity for people to compete with one another for artificially scarce rewards, favors, and positions. This applies as well to departments, divisions, and other groupings. When people are free to cooperate, they will respond with their own intrinsic motivation and will follow the leadership of management who take the trouble to understand a situation and the need for change before making a decision. No company can afford internal competition because it forces people to work strictly for themselves instead of the company.[2]

> ### Axiom 2: Division and Concurrency of Work
> **Taylor**
> *Improvements are due to management's increasing the division of work, and increasing concurrency (different aspects of work being done at the same time), within a project or among projects or processes.*
> **Deming**
> *Improvements are due to increasing division of work, information, and creativity, and to increasing concurrency, within a project or process, or among projects or processes.*

[handwritten margin note: le, k about work]

[handwritten note: not the same]

Anyone who has witnessed a pit stop in high-stakes automobile racing has seen both division and concurrency of labor in action.

Division and concurrency of work have been a driving engine of society. Adam Smith recognized the importance of division of labor and clearly stated that importance in *The Wealth of Nations*. It's the first area discussed in his book: "The greatest improvement in the productive powers of labour, and the greater part of the skill, dexterity, and judgment with which it is anywhere directed, or applied, seem to have been the effects of the division of labour." The parallel concepts of division of labor and specialization are the basis that allowed industry to accumulate capital toward mass production early in the industrial revolution. In 1957, Professor von Mises stated, "The operation of the principle of division of labor and its corollary, cooperation, tends ultimately toward a world-embracing system of production .... International division of labor ... will move forward until it reaches the limits drawn by geography, geology, and climate."[3] We can verify his accuracy by noting the number of products we own that have domestic brand names but were "made in country x of parts made in country y."

Deming's Axiom 2 is not an extension of Taylor's Axiom 2, although the Deming axiom has more depth than Taylor's. Taylor disassociated division of labor from scientific management because he assumed the former as preexisting. The codification of knowledge called for in Taylor's first principle of scientific management was not division of labor; we should not confuse a principle of analysis (Taylor) with a principle of action (Smith). Both Taylor and Deming, however, reinforce a trend that has gone on for thousands of years.

Taylor's studies and principles put their focus entirely on the conventional work processes of his time, ignoring indirect and managerial processes. We now realize that in a Deming-type system (see definition in section 2.1) division and concurrency can be applied not only to these latter indirect forms of work, but also to *information* and *creativity*. Division and concurrency of work bring about improvement through increases in dexterity and productivity. Division and concurrency of information and creativity can provide a second-order effect of improving the delivery of information and tools, which can then be applied back to the division and concurrency of traditional work. It is interesting to look at each of the three.

Division of Work

As people specialize on a part of any type of work, they can improve upon their ability to do that particular task, as working on any one operation or idea is inherently simpler than the total process. Time is saved because people spend less time moving from one operation to another. The overall productivity of the group goes up even more as the various operations are done concurrent with one another. All of these advantages apply equally to machines as well. As operations are isolated by this process, tooling can be used to minimize the amount of human effort required to accomplish tasks.

We should keep separate the notions of, on the one hand, division of labor as an axiomatic principle, and on the other a specific instance of division of an individual's work or assignment. Although Taylor appears to have cared little about the effect on his workers of doing boring, mindless, and totally regulated tasks, Deming's basic belief in the individual's right to take joy in work would preclude assigning people such work on the mere basis of efficiency (such work would also be suboptimizing).

Division of Information

You can divide information, or the people using information, and get more work done. The wide use of computerized database management systems today provides an example. Large bases of data are available as a whole for simultaneous access, with indices and logical relationships differing according to the needs of the various people (or markets or organizations) using them. The development of technology has allowed this to happen on an ever greater basis over time.

Division of Creativity

With the division of creativity, the tools needed to divide and improve work are developed more quickly. The creation of those tools (new products or services) can be divided in order to develop them more quickly. This is especially valuable today because products are more complex as compared to the past. For the creative process to be broken up, first the system and its aim must be commonly understood; next there is definition of the interfaces between components of the system. The interfaces are agreed upon and made common between adjoining components; then the components can be created separately and concurrently, thereby reducing the time required to reach the point of being able to integrate the components.

Additional Benefits of Division of Labor

Division of labor is one of the fundamental phenomena of cooperation in society. One consequence is, of course, an increase of output per unit of labor expended; but the advantages do not stop there.

Historically, division of labor arose from the fact that some jobs are simply too large for the effort of a single person. Other inherent factors, such as the unequal distribution of the physical conditions of production on the surface of the earth, and the inequality among people with respect to their ability to do certain tasks, also played a role.

The system-optimizing nature of division of labor becomes evident when one considers a simple textbook example. Two participants produce the same two commodities, A and B; but one can produce A more efficiently, and the other B more efficiently. The sum of their efforts is maximized when each produces only the product at whose production he or she is the more efficient, rather than each producing both. This is Ricardo's law of comparative cost, described in section 7.1.

Although several more points could be made here, we will confine ourselves to one: thanks to Deming, producers bring not only cost to the bargaining table, but also quality of goods and services—the total cost. Deming's admonition to "end the practice of awarding business on the basis of price tag alone; instead minimize total cost in the long run" is a worthy corollary to both Smith's and Ricardo's laws on the division of labor—and is invaluable advice for any producer who wants to achieve the greatest superiority in production.

Axiom 3: Using Systems

Taylor
Develop systems to perform repetitive tasks.
Deming
Develop systems to perform repetitive tasks.

Whenever you have a repetitive task, you benefit from having a system for describing and doing that task. Taylor developed the idea of using systems to define and design work, thereby articulating an extension to how work is done, moving the chain beyond the point of using interchangeable parts as was discussed earlier.

For Taylor, in his time a system was quite different from what a system is to us now. For example, his view of workers was as bionic machines to do repetitive tasks, and his view of what systems comprise was narrow. In Axiom 4 we'll present another of Taylor's important— and flawed—views of systems. Because of their radically different approaches to systems, this frame of reference is valuable to use in contrasting Taylor and Deming.

The characteristics of developing systems include the following:

The aim of a system must be to benefit all of its parts. Whenever some elements benefit at the expense of others, the system is unstable; it is acting as a transfer agent instead of a creator of wealth. It will eventually run down, disintegrate, or worse be transformed into something that opposes its original aim. This brings out the necessity of having an honorable, and well-stated, aim toward which all the elements of a system can cooperate. As Deming points out in his famous flow diagram (Figure 2), production is viewed as being a system instead of separate, competing

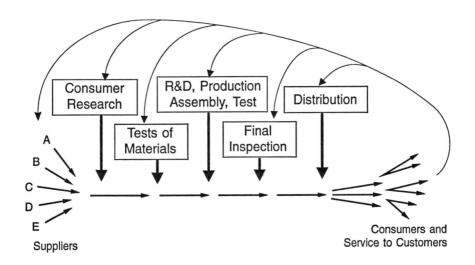

Figure 2. Flow diagram of production viewed as a system.

elements.[4] A business system includes not only internal processes but also its employees, suppliers, customers, and even competitors—elements that both Frederick Taylor and many modern managers treat as outsiders.

Systems are built around repetitive tasks. Whether they take hours or days to complete, are repeated once every millisecond, or cycle only once in a span of years, systems and processes are developed from these repetitive tasks. Repetition is connoted in the meaning of the term system— and as we will see later in section 10.2, "What To Work On," repeated cycles are a key to obtaining the information we need to learn about and improve a system.

Those repetitive tasks have low context level. (Almost all information required for carrying out the task is specified explicitly; little is assumed or implicit.) They are obvious candidates for gaining the benefits of applying the principle of division of labor.

High context level work may also call for specialization, but usually requires an extended period during which, through training and peripheral participation in actual work, one acquires a prereflective grasp of complex situations, a sense of timing, and an ability to improvise in situations not explicitly covered in training. This is the way physicians are trained.[5]

A system can be made for developing processes. Using the science of his time, Taylor realized one could set up a system for developing processes, and that doing so would result in better processes being developed. Taylor's system employed his experts working off-line to create isolated improvements which were then introduced to the line workers. The counterpart of this in Deming's method is the PDSA cycle, which uses the scientific method for never-ending improvement, and involves the knowledge of everyone.

The process of developing systems has changed. As science has progressed, we are in a position today to improve greatly upon Taylor's work. Deming broadened the extent of a system such that it extends at least to the boundary of the organization, if not beyond.6 Thanks to Fisher, Shewhart, and Deming we have the benefit of experimental methods and advanced ways of improving process capability. Taylor clearly provided the extension; the latter improved the methods.

Having developed the perspective of continuously improving systems, we must keep in mind that the people who work within systems need to have *knowledge* of both the systems and the purposes for their creation. This is part of the job of management.

Axiom 4: Optimum Systems

Taylor

The optimum system can be created by proper formulation of the objectives of the system and evaluation of alternatives to meet those objectives. The information will be available to create an optimum system.

Deming

No system is ever truly an optimum system: every system must be analyzed to understand the natural behavior of the system and the variation within it. Information for creating an optimal system is unknown and unknowable.

One of Deming's more nonintuitive precepts is that *all knowledge is prediction*. With the exception of purely historical facts, what we call knowledge at any given moment must remain true into the future in order to have any value we can exploit. Hence Deming's insistence on the predictive aspect of knowledge: knowing something carries a strong connotation of prediction of its continuing future validity. How far into the future? As long as you will need it.

Thanks to the revolution in the philosophy of science that has occurred over the last century—which affected Deming but not Taylor—Walter Shewhart realized that the probabilistic nature of our knowledge about the world limits us to a *relative degree of belief* (that is, a theory) about the future applicability of what we call knowledge today. Hence Profound Change 1 tells us that "the information needed to create optimum systems is unknown and unknowable."

Creating optimum systems is not possible, but all systems—tangible and intangible—can be observed for sequences of variation and behavior. These sequences, which result from an unknown mixture of common and special causes, can be used to gain a relative degree of belief in future performance of these systems; a theory leading to prediction within limits. One of Taylor's fatal assumptions (one that almost all managers still make today) is the absence of common-cause variation in systems, resulting in treating each variation as though it had a single, and knowable, source. (That is one of the two principal cutting points differentiating Taylor from Deming; the second lies in Axiom 5.)

For Deming, the entire organization, including employees, management, customers, and suppliers, comprises a system. This system includes all of the ways in which the elements interact with one another and with the system itself. Management's job is to continually improve the system, and one of their chief means toward this is making cooperation possible. Another is the application of statistical theory and the scientific method.

From our observations of a system, we are in a position to understand whether or not it is in statistical control with respect to some characteristic of interest. We can analyze the system using modern experimental methods to decide what further data should be collected to isolate flaws in the system. This is where Deming has provided the extension beyond Taylor's "one best way," allowing the sources of variability to be treated differently.[7] Yet today practically all efforts still use Taylor's system, or refinements thereof leading to what we saw in rule 4 of Deming's Funnel Experiment: successive application of random forces (tampering) leads not to improvement but to a random walk—*increased* variability resulting in economic loss. A great opportunity lies ahead for those who see the natural extension Deming is providing!

Many people will describe a system using a flow chart or other static-modeling tool, then apply a policy (such as taking out wait-time or cost-added-only processes) to the system as represented by the model. Deming's approach is to analyze the system directly, using data taken directly from the system while it is operating. Too many managers and analysts try to only work from the model, rather than from observation of the actual behavior of the system in operation.[8] Mathematical and other models may possess a certain abstract perfection, but do not necessarily bear any relationship to reality. Learning really comes from the direct observation of the system itself, the "dynamic model"—first, by development of a theory or hypothesis, then from collecting and analyzing the data.

Observing a system in operation, understanding it, and taking intelligent steps to improve it in accordance with its aims is a long-term, and possibly never-ending process. Only management with constancy of purpose will have the vision to keep at this work year after year.

Axiom 5: Finding Causes

Taylor

Once a system has been properly defined and installed, any failure to meet stated objectives must come from outside the system.

Deming

Inconsistencies and contradictions which become apparent upon analysis of the system may be used to detect and isolate the built-in flaws of the system.

The comparison of these axioms brings out the use of the scientific method, introduced by Shewhart and Deming, in order to analyze the data called for by Axiom 4.

Understanding the inconsistencies and contradictions found in the operation of a system gives you the *new knowledge about the system that you didn't know you didn't know*. It helps you understand what other data needs to be obtained. The inconsistencies and contradictions provide clues as to what is not functioning properly. Looking for them was the foundation of so much of Perry's work, because it constantly put him at the edge of the existing paradigm. Rather than ignoring the inconsistencies and contradictions, he used them as a jumping-off point into thinking about why they exist. Doing that allowed him to see the way into new paradigms.

Using the scientific method isolates properties of the system down to a small enough level to identify the flaws. While we will always be denied the perfect information necessary to create the optimum system, we are given the opportunity, through observation and experimentation, to learn more about any system's true capabilities. Since the phenomena we call "inconsistencies and contradictions" are often manifested as nonlinearities in response, advanced analytic tools such as neural networks may be of use in modeling system behavior.

In the process of improving a system using Deming's principles, we are taking out complexity. Complexity has been found to waste as much as 90% of the productive power of a system. Under Taylor's philosophy, we would continue to waste that potential productivity because the idea of any productivity being recovered from within the system does not exist in the Taylor model. When we use Deming's principles, huge amounts of efficiency are available to us that would not have been thought of in Taylor's system. We shall return to this topic in the section "What To Work On."

Axiom 6: Role of Management

Taylor

Continuously monitor the status of the system for deviations from the system objectives to see if improper worker selection, poor motivation, inadequate training, or weak supervision are the causes of missed objectives.

Deming

Create a secure environment so everyone can apply the first five axioms without fear. Offer support, reassurance, and appreciation.

The Deming philosophy lays great stress on continual learning. In our examination of neo-Taylorism we saw that new knowledge is not created by reacting to problems, by aimless measurement, by resolving not to make mistakes, or by punishing those who make them. Axiom 6 points out a basic difference in the role of management between neo-Taylorism and Deming.

On page 38 of *The New Economics*, Deming has a chart for management which shows the big gains are to be found in improving company-wide systems that essentially cannot be measured. Only 3% of the potential gains are from processes that produce figures. Management's faith in empiricism as the basic for improvement must give way to theory—the profound changes and the axioms, and his system of profound knowledge.

Management must see that attention is paid to both the tangible and intangible systems[9] that exist in the organization. A secure environment is essential to carrying out the principles of Deming as described in these axioms. The environment is a key intangible system to recognize and nurture. *Security* is essential so that everyone can spend time seeking and rooting out flaws in the system rather than merely trying to maintain the status quo or, worse, hiding their knowledge of flaws.

Taylor's model concentrates on looking for the mistakes workers make. This model will obviously drive workers toward being insecure, thus keeping "bad" information about the system from management. Workers will compete with one another to avoid blame.

In the Deming system, *management recognize that they are ultimately part of the system*—that they carry the highest responsibility for the productivity of the system as well as its improvement. It is equally important for managers to see how they can gain for the system the greatest leverage by involving and supporting all those who work with the system in the effort to fearlessly find flaws, remove them, and in the process reduce the complexity that robs the system of its ultimate productivity. All the gains achieved in the tangible systems will go for naught unless the intangible systems are improved as well.

The systems that management influence most directly are the intangible ones. These systems must nourish the human factors that bind the organization together and reassure people that their efforts are valued and that they have an impact on the ultimate success of the group. The human factors seen by Deming present a dimension of the system typically ignored by other management philosophies, one that fosters cooperation among all the people in the system. This in turn allows further reduction of complexity, so the division and concurrency of work can be the greatest possible. We will elaborate on this human dimension in section 11.3, "The Restoration of the Individual."

NOTES

1 Charles Hampden-Turner, "The Boundaries of Business—The Cross-Cultural Quagmire," *Harvard Business Review* (September–October 91), p. 96.

2 Alfie Kohn offers excellent insights on the destructive power of competitive behaviors within institutions in his book *No Contest—The Case Against Competition: Why We Lose in Our Race To Win* (Houghton Mifflin, 1986).

3 Ludwig von Mises, *Theory and History: An Interpretation of Social and Economic Evolution*, Yale University Press, 1957, p. 235.

4 The adaptation shown in Figure 2 is almost identical to the one shown on page 3 of Deming's *Elementary Principles of the Statistical Control of Quality* (2nd ed.), published by Nippon Kagaku Gijutsu Renmei, 1952. This volume is essentially a transcript of his now famous lectures to top Japanese management in July 1950.

5 Much could be learned from medical training about how managers might be prepared for their work. An excellent reference on this type of learning is Jean Lave and Etienne Wenger, *Situated Learning: Legitimate Peripheral Participation*, Cambridge University Press, 1991.

6 Deming says, "Industry should also be for the good of the country, not just for profit, managing by outcome." (From the video *A Day with Deming*, Volume 1, George Washington University Continuing Engineering Education Program, 1992.) As we observe elsewhere in the book, this does not mean Deming is against profits or a free market, or that he favors coercion of business toward State-approved goals.

7 A systems-theoretic approach to understanding and improving a system or one of its elements is given in *The Deming Library* series, Volume 21, *A Theory of a System for Educators and Managers*. Dr. Deming and Dr. Russell Ackoff are the speakers.

8 For a welcome shift in motive in the use of models, see "Modeling to Predict Or to Learn?" by A. P. de Geus of the London Business School, in *European Journal of Operational Research* (59), 1992, pp. 1–5.

9 These are discussed in section 10.1, "Types of Systems Within an Organization."

Deming's Profound Changes to Management

Up to this point, we have developed a foundation based upon understanding Deming's principles of quality and productivity improvement. Now we shall look at those to whom this base of theory is directed. We very willingly agree with the stand that everyone must be involved in the transformation, but management occupies a special place in carrying out the implementation.

It is likely the reader has developed a sense along the way of the role management plays in shaping the processes that make up the overall system of the organization itself. Clearly management has ultimate responsibility for both the system of the organization and its outcome. Management also bears the real power and authority to make changes in that system happen. And, although the authority can be distributed throughout the organization to leverage all the resources available, management still retain the *final* responsibility for the success of the system they create and foster through their decisions and actions.

The next few sections provide some thoughts for consideration of what it takes in the motivations and skills of a manager to be a truly effective leader of change for the organization, regardless of perceived personal style.

9.1 Why Do People Want to Be Managers?

In the first five years of Perry's consulting practice, he noticed a significant increase in the number of hours managers were expected to work, and they were being pressured to work more. Somewhere around 1984, he started wondering what advantages being a manager had when one started having to work *more* as well as harder. The results of this question helped him understand more about leadership.

In typical American companies, most people want to be mangers for some combination of motives that could be grouped under the categories of "altruistic" and "self-serving." Generally people continue in management because it gives them satisfaction and access to implementation of ideas and changes—regardless of whether the principal driver exists in the "altruistic" or "self-serving" category. In some cases, however, managers continue in their jobs when they are no longer satisfied, due to a sense of duty, desire to maintain a standard of living, or fear of inability to perform in other jobs.

Altruistic	**Self-Serving**
• Responsibility/Duty • Volunteer • Election by Others • Ability to Implement Ideas and Changes • Satisfaction • Natural Leadership/Talent	• Power/Influence • Recognition • Standard of Living • Ability to Implement Ideas and Changes • Satisfaction

Table 7. Some Motives for Becoming a Manager

In the "altruistic" category, everyone has seen those leaders who volunteer or are elected by others to serve as managers—the job has to be done, so they do it, often with seeming ease. Managers so motivated tend to exhibit the characteristics of flexibility and adaptability so necessary for survival of the organization.

The "self-serving" category can involve election by peers or subordinates and/or more senior management. Regardless of the selection process, a sense of duty/responsibility or a desire for power, recognition, influence, or increased wealth drive the individual. The reward mechanism, in addition to its appeal to monetary gain, becomes centered on the individual's sense of satisfaction, which is mostly derived through a manager's ability to effect change more and more broadly in the organization. Managers who are motivated by power, influence, or recognition tend to develop a rigidity of style that will negatively affect the survival of the organization. The larger the organization, the more pronounced this tendency will be.

Western business is in decline today, with many companies whose names were household words disappearing or going out of business.[1] If you have any special knowledge of any of these companies, ask yourself which of the two families of reasons listed above motivated its management.

We must recognize that changing an organization is best accomplished after considered study, which often lasts for years. Changes made by a new manager for change's sake or for recognition are usually detrimental in the long term for the organization; they cause uncertainty, doubt, fear, and competition among the employees, who develop ways to minimize the negative impact of the change on themselves.

The question we must address here is how to get managers to accept and make *profound* changes. Because one of the basic tenets of Deming's profound changes is to reduce complexity, it is easily understood that you cannot expect those managers driven by power, influence, and recognition to welcome Deming's changes. Complexity is often used by these managers to secure their position—yet, paradoxically, the higher the degree of complexity they cause, the more they feel the need to carry the load themselves.

We believe in four ways to develop managers who will lead and support profound changes:

1. Give people better training in the philosophy, theory, and actions needed for profound change (transformation).
2. Select people for management who know, understand, and practice Deming's teachings.

Notes for this chapter are on page 185.

3. Inspire self-esteem, which thrives on the profound changes and trust/empowerment.

4. Provide support for longevity in management jobs to discourage the job-hopping prevalent in U.S. industry.

The responsibilities of managers are developing people, providing an effective organization, and setting direction through appropriate decision-making in technology and the business. We must also understand that leadership is not the sole possession of management. Most successful organizations have strong leadership, all the way from the technical community to the factory floor in partnership with management. The effective, entrusted organization will display teamwork and individual leadership throughout, regardless of the experience and titles of the people involved.

Managers who are continuing to overstress and overwork themselves and their organizations should confront themselves with the dichotomy in Table 7 and be honest with themselves as to where they stand.

9.2 A Manager's Options

**Give Up Power and Authority
(Rejoin the Workers)**

——

Learn New Skills

——

Retire Early

——

Die Young

Whenever we try to start quality, productivity, or any other sort of improvement programs managers are always concerned about their own options: what will the program mean for their career and their supervisory position? Unless they get the answers that really satisfy their need to maintain and advance their careers, they will probably pursue some other program that promises them better options.

Interestingly, managers will even opt for programs they know are definitely not in the best interest of the company. The important thing for them is their own options, even at the expense of the company. After some years we now understand the situation well enough that we can give managers a set of options that are in their own long-term self-interest as well as in the interest of the company.

All things considered, rejoining the workers is a viable option for a manager because our experience has been that workers undergo less stress than do managers. If they want to remain managers, they can learn the new management skills (leadership, communication, and cooperation) needed to make the profound changes.

Early retirement is one option being selected often. In some cases managers are being pushed out by their company when they are in their fifties, but we are also seeing managers retire when they are in their early forties and sometimes late thirties.

If managers don't like the other three options and making the profound changes is still not attractive, then they can try to do the work of the system: personally doing administrative tasks as staff is cut, creating plans and reports without asking if there continues to be a need for them, and trying to maintain an image of improvement when the system makes improvement impossible. The resulting stress may give them another option—dying young.

As we have seen earlier, pushing on the processes to get around system defects will not help us be competitive. If anything, the pressure felt by people throughout the organization to do what they cannot do raises their level of stress. That stress can then lead to negative behavior being fed back into the processes, causing poorer results and lowering the value of the product or service offered.

People are asking about their options, and we have listed some of them. People suffer physically and emotionally when they do the work of the system rather than working on the system to reduce complexity. This is very evident in managers, but in general this applies to everybody. Instead of reducing complexity and allowing the results to change the system, management today are preoccupied with attempting to work directly on the system by flattening the organization and downsizing—which increase the stress levels of managers, who, if they don't retire early, often die young.

9.3 New Skills for Managers

```
LEADERSHIP

COMMUNICATION

COOPERATION
```

As noted earlier in this chapter, leadership, communication, and cooperation are the personal skills that managers must have and continuously improve in order for their organizations to be successful in the competitive global market. These skills are critical to providing direction and effectively utilizing resources.

Management have the responsibility for creating and improving organizational systems. That responsibility simply cannot be undertaken without these essential talents. The following discussion goes into further detail on each of them.

New Skills: Leadership

Leadership can be defined as providing people with a picture of what needs to be done to achieve common objectives, and instilling the desire to achieve them chiefly by actions rather than by rhetoric. Most effective actions occur after clear visualization is provided of the results expected to occur from those actions. Deming's system-flow diagram would aid the visiualization process. Deming would call this developing constancy of purpose. That *vision* ultimately works with the intrinsic ambitions of individuals to allow them to do their work well and to add value.

In order for a group to perform well, *trust* must exist among all the members. Trust leads to collaboration instead of competition. It leads to the growth of a learning environment where faults are treasured, instead of an internally competitive culture where even pointing out faults can be grounds for expulsion from the group. Leadership actively encourages behaviors that allow trust to exist and grow among all group members—and that absolutely requires leaders themselves to be models of the desired behaviors.

Providing the picture of what needs to be done is only one example of the need for managers to lead by *communicating openly and frequently* with the group and all those with whom the group interacts. Management's typical role as a "reality-distortion field," as one of our students once put it, must give way to a solid flow of true information to and among the members of the group and in both directions with the rest of the system. And while encouraging open communication is important, it should also be performed thoughtfully and *carefully*. It should be constructive and understood to represent the truth. Communication to others should be of the highest possible integrity. Poor communication creates uncertainty and fear that in turn increase variability, kill trust, and reduce creativity and productivity.

True leaders constantly seek to *understand* the impact they and their group have on the rest of the organizational system. They will probe without being evaluative, and seek clarity when there is the potential for misunderstanding. Leaders will check to see that their efforts, and those of their team, best support the ultimate results of the full organization. "Best support" may sometimes mean that the ideal results of the leader's immediate group must be foregone so the results of the full organization can be improved. This *cooperation* will exist in a constant cycle of better understanding how the results of the organization are being reached and what is needed within the organizational systems to achieve them.

It is necessary that the manager of a group not be the only one who displays leadership. As much as possible, leadership should also come from the other members of the group. One way to encourage this is for the manager to teach leadership by example. When leadership does arise within the group, the manager should encourage the effort and provide it with the support and resources that will assure the group's continual progress and development. This ultimately leads to the manager *entrusting responsibility and accountability* among all members of the group, developing a division of labor and sense of ownership. This is in contrast to the standard Western management habits of giving directives and using MBO (management by objective) to "ensure" compliance.

The skill and responsibility of leadership will also find the manager being personally involved in the training and education of others in the organization. This will include involvement with the design of curricula, plus research and presentation of material. This helps assure that the individual who best understands the needs of the organization is directly involved in the development necessary to carry out the organization's goals. This isn't to say that the manager will not be assisted by others—he or she most certainly will be one of a team of people chartered to identify and provide the necessary training and education found to be required. Ultimately, all members of the team will develop particular areas of expertise where they are the most knowledgeable and will be called upon to help in the effort to educate the rest of the team. But the manager who is a *leader* will be involved in ongoing training and education as the needs of the organization are determined, be it as a coach for an individual or in a more formal classroom grouping.

Finally, a leader is not a barrier to continuous improvement; and a leader makes clear to the organization that all such barriers will be removed. The real leader will *facilitate the processes of the organization* according to Deming's principles, using the above as guidelines in order that the organization can best achieve its vision.

New Skills: Communication

In communication, both actions and words are important. Managers usually have some awareness of this, but often tend to fall into a false sense of believing they "know it all" or otherwise feel the techniques they have successfully used in general experience will be appropriate in all new settings. They might also use this communication experience only to forward their own personal ambitions, unaware of the opportunities to work within and really understand the organizational system in ways that can help the organization achieve higher levels of performance.

The new skill of communication requires that the manager appreciate the need to *continuously listen, to use both eyes and ears.* A manager's eyes and ears should provide the information on how others in the organization act and communicate with one another. Providing that a manager can keep ego from getting in the way of hearing, the manager can detect the various cultures within an organization.

All organizations have cultures within them—groupings of people with shared beliefs, behaviors, and language. They can be found by observing how people interact with one another to get things done together. Organizational cultures are pervasive and will be an inevitable factor in how communications flow into and across the enterprise.

People within these cultures have noticeable ways of acting toward and speaking with one another. Those behaviors and language will be different as one goes from one part of the organization to another. The cultural differences can be simple and straightforward (phrases, style of dress) or complex, yet subtle (patterns of behavior, inflections, context levels).

When communicating to others, *knowing and respecting your listener's culture* should be taken into consideration. Managers should learn to be keenly aware that the farther a non-manager "listener" is from the manager "source" (in an organization, "distance" means levels of hierarchy as well as physical distance), the more out of touch the listener will feel.

Within just a single office there will be cultural differences as people go even a mere fifty feet from their own work areas into other departments.

One complexity of cultural elements is demonstrated in the existence of *action chains*. Edward T. Hall defines an action chain as "a set sequence of events in which two or more individuals participate. It is reminiscent of a dance that is used as a means of reaching a common goal that can be reached only after, and not before, each link in the chain has been forged. . . . If any of the basic acts are left out or are too greatly distorted, the action must be started all over again."[2] Once a particular sequence of behaviors (actions) connected to a dialogue is started, acting in a way not established for the chain may result in terminating effective exchange. A simple example might be how some people commonly start a phone conversation with "How have you been doing?" When that preface to the conversation is omitted, the uninitiated speaker may be perceived to be unfriendly and potentially threatening. Another context for action chains is in the receiving of gifts. In America it is considered extremely bad manners not to acknowledge a gift: the more special the gift, the more formal the acknowledgment. Breaking this action chain by not telephoning or sending a note is confusing to the giver and may result in fewer, or less thoughtful, gifts in future.

An example of a rather complex action chain is how two people are introduced to one another. Although an introduction may be concluded in a few seconds, it is a very complex action chain affected by the sex and age of the participants, and by who is doing the introducing, whether a handshake is included, and what interaction is expected to follow the introduction. The handshake itself is an action chain involving eye contact, degree of firmness, duration, and what the parties say to one another.

The downside of action chains is that individuals not familiar with the accepted chains in a group's culture can unwittingly cut off dialogue and become perceived as "not communicative." This indicates that *managers need to understand the action chains their listeners expect and respond appropriately*. They have to be sensitive to introducing new "buzz words" or developing new sets of action chains without first determining how they can be correlated with the existing underlying culture.

David Packard of Hewlett-Packard was a good example of an effective leader who practiced the approach of understanding the language and action chains used in an organization's culture. Even as the "big boss," he saw the need to be personally involved in communication at all levels of the company, in ways that were sensitive to what the audience was comfortable with. Over the many years of growth in HP, he often met with employees, not only in scheduled "communication sessions" but also in less formal after-hours settings (both at company-sponsored "beer busts" and non-official gatherings away from the workplace), relating to them at their level in their territory.

In conclusion, what's increasingly important in communication is to understand the culture in which you are trying to communicate, and to be aware of the language and action chains that are used. Here, we're talking about culture at a very fundamental level: as found, for example, in differences that will probably always exist between engineers and sales people; between day and swing shifts; between the lowest clerks and the highest executives. Even in an organization that is transforming to Deming's principles, these factors continue to exist.

There is more to organizational culture than using the same process terminology and solving problems the same way—though those few factors are often the only ones emphasized in popular management training and practice. As stated above, a good communicator will take the additional elements into account, and will use an awareness of the fundamental cultural factors of a group to talk and listen actively in ways that are most acceptable to that group.

New Skills: Cooperation

Cooperation deserves to stand out as a particularly necessary factor for management to develop in any organization looking toward long-term success. We need to identify encouraging *cooperation* as a fundamental skill requirement and define the skill. It is a skill that is clear and distinct from leadership.

Actually, cooperation is the fundamental *result* of understanding the situation and the need for change or more data *before* making a decision. Taking this perspective ultimately takes the manager of the organization from being subjective (and, more often than not, tampering with the system instead of gaining long-term improvement), to being data-oriented and managing by facts.

Of course, managers may force people to cooperate—or more accurately comply over a short term—but true cooperation can only result when a group of people effectively work together toward a goal they truly share in common, where their ultimate gain is seen as a gain by the whole group. Cooperation does not exist when individuals are focused on their personal gain from whatever advantage they can achieve over others within the group. Enlightened management will work to remove the obstacles that foster individual personal gain over collaboration. Some examples of those barriers include awarding "Top 10" performance bonuses, instituting merit pay (forced ranking) systems, determining benefit allocation by level of job grade, personalizing problems to individuals or groups versus looking at flaws in the system, having all reports flow only along the lines of the formal organization chart, and so on.

One is reminded of those situations where everyone in an organization is aware of a crisis being at hand except the manager until he or she is finally presented with the data. Opinions typically abound, but only data can provide the information needed to truly understand what is taking place in order to develop a solution which will seek to eliminate causes and provide lasting value. To gain cooperation managers must work toward using and providing data, not opinions. If not enough data are provided or available to make the picture clear, managers must take the time to encourage the various constituents to work together to better define the issue at hand and obtain more data until it is clear what action is necessary. As success is achieved through the work becoming easier, the parties will come to appreciate the benefit cooperation provided them. Future situations will find the participants more likely to seek one another out based on their past mutual success, and a positively reinforced cycle of cooperation and group learning will develop as a result.

9.4 Leadership

Leadership states what is expected of the organization.

Leadership envisions what the organization will try to do (the aims of the system) and what it will not try to do.

Leadership works to make the means available for the organization.

Leadership understands that not all needed actions are knowable or feasible due to variation in the system.

We have shown leadership's bottom line. Leadership needs to pursue this bottom line if it is to be taken seriously.

All organizations, regardless of size, need to have someone thinking about the organization itself. In Deming's philosophy this is clearly part of top management's responsibility; only the leadership has the requisite authority, vision, and constancy of purpose. The leaders must describe the limitations and expectations of the organization, along with how it interacts with the other parts of society.

Throughout, the leader understands that natural variation within the system will mean that all the knowledge required to operate ideally does not exist.

NOTES

1 In many cases a declining company's trademark, name, or logo is bought and used by another company. Our interest here is in the dying organizations themselves rather than physical manifestations such as names or brands.

2 *Beyond Culture*, Anchor Books, p. 141.

Understanding and Improving Systems

We need a system to gain the advantages of division and concurrency of labor.

We need a system to ensure enduring contribution by the organization.

Management's role is to see that appropriate systems are established, supported, and improved.

Not having a system is like doing every job for the first time, every time—including the accompanying errors and false starts. Without the regularity and repetition of work that characterize a system, it becomes much more difficult to learn because we are deprived of time-ordered process data. Without a system, workers, teams, and departments would act independently of one another, without synchronization or division of labor.

If we have no system we have no opportunities for progress and productivity improvements.

One of Taylor's main contributions was to recognize that if we have a repetitive activity (or can organize it so it can be repetitive) we have the opportunity to create a system. Not doing this where possible is throwing away a major opportunity. Having a system allows us to refine the operation of the system in accordance with its aim as we observe it over time.

10.1 Types of Systems Within an Organization

Most people understand that their organizations have a multiplicity of systems, but few realize there are actually three different *kinds* of systems. Improvements confined to one or two of these three kinds of systems are unlikely to have any lasting good effect.

Formal Systems

Formal systems are typified by written documentation which is often handed out to people. Reorganizations work at the formal system level. Personnel, auditing, and purchasing, for example, are usually quite formalized systems, and manufacturing processes are often formal as well. Relations with suppliers, and indeed with employees, are usually prescribed as formal systems. In general, formal systems in an organization are either well known or the documentation can easily be obtained.

Informal Systems

Informal systems are the "shadow organizations" that grow up around and within formal systems as ways of actually producing required outputs. Informal systems are evidence that formal systems either don't exist, are too difficult to follow, or don't produce the desired results. Informal systems involve networks of people throughout the organization who cooperate informally (sometimes outside their actual job responsibilities) to get things done. Most people learn how to do their jobs by observing how the informal systems work.[1]

If it were not for informal systems, most organizations would perform little or no work. Reorganizations may or may not affect informal systems. Managers sometimes rely heavily—and unrealistically—on the continued integrity of the informal system as they plan formal reorganizations that look good but actually increase complexity.

Intangible Systems

Intangible systems comprise the knowledge and lore of the organization that tell people their limitations and where they really stand. All companies have intangible systems. The management transforma-

Notes for this chapter are on page 203.

tion Deming calls for will have its greatest effect on the intangible systems. Fear, which many companies apply deliberately to their employees, is an example of a part of the intangible system; it forms an invisible barrier to quality, productivity, and creativity.

Employees may experience this fear but be unable to identify or prove its source. We have found it is possible to find certain repetitive events in the intangible system that promote fear, such as quota-setting and performance appraisals. By collecting data using the "What's bugging you?" technique, referred to in the next section, we can proceed to find and isolate the sources of the fear, thereby leading us to what to do to reduce it in a way that objectifies people's local, personal knowledge.

A company may have a *formal* grievance process and may ensure that each employee has a printed copy of its documentation. From experience in the organization the employee learns the *informal* system which actually handles such grievances, and that system may vary considerably from the formal process. The lore of the organization in its *intangible* system may tell the employee that submitting a grievance may not be good for longevity, despite formal assurances of security.

Regardless of type, whenever any system guarantees or demands predictable results every time certain inputs are applied, it is neo-Tayloristic. Instances of this abound in every organization. Examples would include management by objective (MBO), demands for continuation of trends, guarantees by consultants that people trained in their methods will achieve a predetermined level of improvement, and certification by quality organizations.

10.2 What to Work On[2]

Early in my consulting career I noticed that organizations waste huge amounts of time deciding what to work on, and they often pick issues so deep that they waste even more time trying futilely to work on these.

Sometimes the organizations would become so frustrated with

lack of improvement that they would change approaches, and sometimes change their consultant. Some wasted time by adopting some fad such as quality circles, and training everybody, only to be amazed later when no progress occurred. Others would pick the same problem to work on as was talked about by some CEO in a magazine article, not realizing that common, or stable-system, variation *alone* could account for the apparent success of the other company.

In general, I have found it takes only a few minutes to figure out what to work on. The items picked can usually be worked on fairly easily and with a very profitable return on the effort.

One of the key things I learned is that all systems have *constraints and bottlenecks*; and a lot of time and energy can be saved by focusing on bottlenecks, the things that limit the system from achieving higher performance. It is impractical and undesirable to try to eliminate *all* bottlenecks, hoping to get a system with balanced capacity, because the sequential variability in any system will just make the whole system a bottleneck. *Instead, use bottlenecks to control the scheduling of your system, and make sure any system improvement is applied at the bottlenecks.* (And recognize that bottlenecks may move around.)

Often groups who try to copy "success stories" are unable to reproduce the results because what some other group worked on was not a bottleneck for them.[3] In some cases, people will spend time diagramming the formal system looking for a bottleneck, but the way I'll discuss is much easier and quicker.

Another thing of key importance is that business issues can be classified very nicely into three important groups: **Cosmic Issues, Low-Hanging Fruit, and No-Brainers.** Many efforts have failed because people tried to go after cosmic issues directly. *Cosmic issues* are too hard to solve directly because they deal with the health of the entire organization, require vision to tackle, and are highly complex.

Sometimes people would tell me that the examples Deming shows in his books and classes to illustrate improvement looked like no-brainers. This was because the work had already been done on each problem;

all we were seeing was the result, which made the whole example seem a lot easier than it had to the people who had to do the original work to improve the system. *No-brainer* is a description we apply to the problem *after* we have solved it; whether something is a no-brainer is difficult to assess beforehand. If you keep trying to find no-brainers, you may lose the impetus for continuous improvement.[4]

Finding the right level issue to work on is like the problem Goldilocks had with the Three Bears' bowls of porridge: we want to work on the problem that is "just right" in terms of level. This is what I call *low-hanging fruit*: you can just reach it, but it takes some effort. Table 8 is a guide that will help you decide which of these three categories the issue you are considering is in.

Experience has shown that it is a long, arduous process with uncertain success if you work directly on cosmic issues. It may seem profitable at first to go directly after no-brainers, but we have found it is better to restrict ourselves to low-hanging fruit. The no-brainers come out of the woodwork as we work on low-hanging fruit, and give us a handsome return for our effort. So our strategy has been to pick low-hanging fruit, then to pick off the no-brainers that are available. People ask, "Don't we have to solve the cosmic issue anyway?" The answer is that after we solve enough no-brainers the underlying cosmic issues tend to go away along with the low-hanging fruit.

This strategy has been so successful that we have only had to use the PDSA cycle (Shewhart cycle) on no-brainers, and only a couple of times have we had to use it on low-hanging fruit; yet productivity has increased two to three times with the same people and equipment. The philosophy here is to work as though peeling an onion, working inward on problems by starting from an outer layer.

For years after Deming announced his Fourteen Obligations of Top Management, most people thought they were the agenda that we were to work on. The Fourteen Points were never intended to be treated as a to-do list. Using them in that way, especially with expectations of perfecting your system, would be Taylorism, not Deming.

	Cosmic Issue	Low-Hanging Fruit	No-Brainer
Time Frame	Long	Medium	Short
Measurements, Data	Unknown	Measurements known, data unavailable	Data available
Consensus, Awareness	Low	Moderate	High
Fear	High	Moderate	Some
Development of Trust	Difficult	Moderately difficult	Easy
Return on Investment	Unknown and unknowable	Fuzzy	Certain
Resources Needed To Solve	Unknown	Uncertain but available	Clear
Development of Control	Difficult	Easy	Total
Status of Problem and Solution	Unknown	Problem clear, solution unknown	Clear
Change in Direction	Major	Minor	Minimal
Division of Issue	2 or more low-hanging fruit	2 or more no-brainers	None
Interdependence	High	Moderate	Low

Table 8. Characteristics of Business Issues

A lot of improvement work today concentrates on the *formal* system. The usual procedure is go over the formal system in great detail and look for places where there is no or low value added. But this depends on having your formal system well documented—most companies don't, so they wind up spending time documenting the formal system. But formally documented systems rarely behave like their documentation, so even more time and effort are wasted.[5]

In addition, we have found considerable advantage in working on problems that are in the *informal* system, which by definition is not formalized, and in the *intangible* system, which can't be formalized (see section 10.1). Some companies leave many processes undefined, and trust the education they give their people will be enough in the hands of sensible employees; this is one example of an informal system.

We have found a quick way to identify low-hanging fruit: a simple procedure that we call "What's bugging you?" This process may seem at first like another way of describing the PDSA cycle, but it is what we do first *before* we use the Shewhart cycle.

The first step is to ask ourselves the question, "What's bugging you?" about our work. We could also ask our customers what's bugging *them* about our work. It's very important that the responses to this question identify low-hanging fruit (see the Three Bears problem mentioned above). We then take the most important two or three of these issues, and list the common factors among them.

The next step is to find out *what is repetitive* about these issues and factors: what is it that happens again and again, and where in the process does it happen? This will help us in deciding what and where to measure.

The important thing so far is that you have come up with a *simple way for people to share their subjective and local knowledge of the problems in the system.* The goal is to go through a couple more steps which confirm your knowledge, so it is no longer subjective and local but is objective and global.

Now we are in a position to use data to acquire knowledge about the process. We describe how we would measure the process to learn about it, and *then we go after the data.* In the best case, we would find that the data already exist, which costs us almost nothing.

We are looking for data that will show us variation over time for one person, group, or machine; or variation from item to item, machine to machine, group to group, employee to employee. By taking measurements we are in effect making physical processes into data-generating processes which allow us to observe them and learn from them. An essential thing to learn about the process is whether it is in *statistical control,* which we can tell from plotting the data on a control chart. If the process is not in statistical control, we try to analyze why by looking for and removing the causes for the out-of-control points. If the process is in statistical control, we start looking for subgroupings of the data which we can subject to additional analysis.

At this point it's important to realize our theory is that all points on the chart are caused by the same set of random causes; that is, we see no reason yet to assert that any special groupings exist within the data which might have different cause systems. Our hope, however, is that this is an incorrect theory; and we are hoping to find inconsistencies and contradictions to it in some groups of data. The tools we use would be Pareto charts, scatter diagrams and control charts, analysis of variance, and even designed experiments. If we do find an inconsistency or a contradiction in our theory, we have detected a flaw in the process; if we can find none, we have to refine our ideas of how to measure the process.

A *verified process flaw* is very important because we can be fairly certain it has a common cause. In order to help us find where to change the process, we have to get greater clarification as to the source of the flaw. This will probably require additional data and will be used to probe again for inconsistencies and contradictions to our theory.

Although this analysis sounds time-consuming and expensive, we have actually been working fairly fast and cheaply, especially when compared to what comes next. It's because of the expensive nature of the next steps that we've put so much emphasis on finding where it is easiest to correct the process, and where it will have the most effect.

The categories of cosmic issue, low-hanging fruit, and no-brainer, despite their everyday names, are theoretical constructs, although the advice to concentrate on low-hanging fruit is empirical. Once the problem is picked, the approach to solving it by modeling its repetitive nature is based on theory developed by Shewhart and scientists before him.

The next step is to apply the Shewhart cycle. The benefit of beginning with our "What's bugging you?" analysis is that we can now start the PDSA cycle with a good idea of what to change in the process. Our idea is based on *theory and observation*—rather than simply acquired and acted on randomly.

Correcting the system is essential. Pointing out flaws in the system and simply removing defects from the product are not enough. In addition, the above analytic process must be applied in the context of a holistic, system-wide viewpoint in order to avoid suboptimization. Even if each component part were made perfect it, would not follow that the system is perfect; the system is synergistically more than the sum of its parts.

10.3 Reducing Complexity in Product Development

In product development every product is different. The one thing that remains the same is scheduling. *Scheduling* is the tool people use to control their time and the other various resources that go into the project. The repetitive structure of scheduling gives us a system that can be analyzed—adherence to the schedule, integrity of the schedule, availability of resources, and accuracy of completion estimates. The emphasis here is that repetitive aspects of the development process can be thought of as a system in order to improve it—and we can analyze these aspects in order to improve the overall process. The only challenge is to be creative in our thinking about development as a system and patient in collecting the data.[6]

Restricting Flow at Bottlenecks

Management, who don't like to see high-priced people sitting around, often insist on having them work on many things at the same time, which stuffs the system with work. But the bottlenecks still limit what can be done. Since no one wants to work on a project destined to be scrapped, factionalization occurs when people defend their own projects. If we restrict projects to what the system (that is, the bottlenecks) can handle, this excuse for factionalization goes away.

Bottlenecks can be used as scheduling devices. First, we recognize and identify them as bottlenecks, and then we do not do anything to enlarge their throughput, such as creating parallel paths or upgrading equipment. Restricting the allocation of funds and personnel could act as a deliberate barrier to tampering with the bottleneck *at this time*. It's not that we believe the bottleneck should never be moved, but that movement should be done only after great thought—and never as the first step in reducing complexity. When the bottleneck is opened up, we must always understand that another one will show elsewhere.

Another device Perry Gluckman used to control chaotic changes to his clients' schedules, which were causing ripple effects up and down the system, was to institute what he called "red" and "green" days. On green days the schedule could not be changed and was carried out as planned; on red days they admitted the chaos and allowed the schedule to be changed. As far as the scheduling system went, red days were the only days on which they were allowed to have problems. On green days everyone knew the schedule would not change and could spend the time executing it. Perry also used green days to analyze why activities due that day were not completed; often the percentage of activities completed on time rose from as low as 60% to 100%.

Organizing Chaos

Is the system of scheduling stable or chaotic? We can systematically use Gantt or PERT charts to analyze the transition from one activity to another: was it on time, were the needed resources available, did something unplanned occur to cause delay? Typically, people will pad their schedules to compensate for all sorts of uncertainty, so this analysis can lead to reduced cycle time. Care should be exercised in the analysis itself so that people do not feel discredited or threatened—it's the *process* that should get the critical attention. The analysis should lead to identification and removal of the causes of problems so the uncertainty can be decreased. If the only action taken is to admonish developers to work harder, the chance of reducing any of the disorder is nil.

Reducing Working Around Missing Resources

As we have found in production operations, management in product development have just as much tendency to work around missing resources. Some examples of the resources that can be missing in product development include knowledge of the market, clear product strategies, technical specifications, technical knowledge, operational definitions, funds, equipment and tools, people, and training. Similar to the earlier discussion in the section "How Management Promotes Complexity," attempting to work around missing resources produces more rework as design definitions change, as discoveries are made of what customers *really* need, as two groups work independently without a common and comprehensive interface definition only to find their modules are not compatible with one another, and so on. Our advice is to proceed with no more work on a product until the missing resources are provided. Put the labor available into projects where no resources are missing—those projects will finish more quickly, and the projects with missing resources can be focused on once the needed resources are made available.

Design and Redesign

Deming's system flow diagram begins with what he calls Stage 0: the generation of ideas. This is where the design for a new product or service is originated, or where ideas for improvement bear fruit in redesign. Keep in mind that the producer, not the consumer, is responsible for quality: you must find out what is needed and anticipate it. During this period it may not look like anything is happening, but quality can be no better than the design.

Reducing Tampering With Specifications and Design

Take the time at the beginning of the project to specify fully what is being developed. Then stick to that specification so the product can be released much sooner. If people are allowed to tamper with the specifications throughout the project, this causes rework by the engineers. Consider developing less function into the initial product, getting it released, and then adding function after having customers work with it. In software creation, for example, it's been observed that roughly 80% of the users of a new program or system use only 20% of its available functions. Since most software has a lifetime of several years, its maintenance and enhancement costs can be reduced by producing a small core system as the first version, and by only adding those features, from time to time, that users really need. Without this approach, the software's developers will tend to expand complexity beyond their own ability to implement, and will be maintaining a large amount of function that is seldom used. Involve the support groups and operations people *in the beginning* who will service and produce the product—getting their input to specifications will also help reduce or eliminate the rework typically done when projects don't involve the other groups until late in the cycle.

10.4 Material and Resources Flow

For more than a decade, manufacturing organizations have worked to shorten cycle time in their flows in an effort to achieve a just-in-time (JIT) flow of materials and resources. The practice has been refined to its greatest extent by Japanese management, aided by their understanding of Deming's profound changes. Regardless of nationality, several organizations have had some measure of success, yet many are still struggling to get consistently good results.

The key to reducing cycle time successfully in the flow of materials and resources lies in reducing the overall complexity within the system of that flow. Simply "implementing JIT" by having suppliers deliver goods frequently in small batches and reducing planned production quantities to units of one will not alone obtain the results for which you hoped.

The complexity to reduce in the manufacturing system will exist in more forms than can possibly be covered here, but managers and planners implementing JIT have to appreciate that the sources of complexity in the system will be tangible and intangible, formal and informal. The informal and intangible factors (such as how the individual members of the work group personally recognize one another's contributions and what memories of past occasions they use to reinforce their beliefs they are doing "the right thing") will need study. Dealing with the formal and tangible factors is important, but not paying attention to the informal and intangible factors will just as certainly hinder the success of improving the flow of material and resources.

Everyone involved with the flow will need to be educated in the plans to streamline it. It will be important to give all members of the group a voice in how the plans are implemented. During and after the initial implementation, team (functional and cross-functional) review of both the process and results will need to occur periodically. Department boundaries and lines of communication may have to be changed, so the traditional barriers (such as those which often exist between planning and production) can be eliminated. These are only a few examples of the many that will have to be addressed.

Also review the sections on complexity and consider how it applies. For instance, the more steps there are in the flow, the more complexity there will be—benefit will come from removing steps that don't add real value. In the Red Bead Demonstration almost the entire process is determined by the built-in complexity.

The physical flow of materials is not the only formal process to pay attention to. For instance, the flow is driven by the processes of forecasting and planning. Process technology (the technology of systems thinking, studying variation, systematically reducing complexity—not the technology of computers or spreadsheets) can be used for this as well as for the physical flow. There will be variation in forecasts and the production planning driven by those forecasts. Some of the causes will be special; others will be common to the processes. Using statistical tools, the causes of forecast and planning variation can be studied and steps can be taken to eliminate many sources of variation (such as tampering by management through sales incentives or padding of the schedule), and other sources of variation can be minimized.

Further complexity can be removed from the flow of product by extending efforts out to suppliers. Often the most effective method is to share Material Requirements Plans (MRPs) with them. An important factor in doing this: the supplier must be made aware of the factors that influence the customer's business, and must have assurance that variations in production (as when the MRP forces more build than actual demand calls for) will be ultimately absorbed by the customer.

It is also important to determine what the bottlenecks are in the flow. As is explained in the previous section, the bottleneck in a process can be used for scheduling what routinely goes into the flow. Bottlenecks can be used as important tools in planning, but they must *not* be allowed to move around. When independent efforts are going on to optimize portions of the flow, the point of greatest flow restriction will change. Moving bottlenecks around in this way will negatively affect the ability to systematically plan for total output.

When the combination of above factors are systematically taken into consideration, significant decreases can be made in the time it takes for materials and resources to move through a process. We have seen the direct result of the application of these principles bring about decreases in the range of as much as 90 to 95% of the cycle times the processes had before the complexity was removed. But these suggestions are hardly exhaustive. Consult "What To Work On" earlier in this chapter.

10.5 Hierarchy of Opportunities

When starting to work on improving systems, most people will likely start with production-type processes. In dealing with the first four levels of hierarchy, work can be done on both the "hard" and "soft" aspects, where *hard* refers to the product or service itself and "soft" refers to the processes that make or control the product or service.

Table 9 provides some examples within this model of hierarchy, going from higher to lower, but these are just a few examples of the hard and soft aspects of products and services in a hierarchy. In his book *Everyday Heroes*, Perry tried to show a number of examples of hard and soft factors related to quality, and how they can merge together at times.

Looking at this concept of hierarchy, one can see many situations where a process output appears to be out of control but really isn't when viewed on a wider scale. Common causes at a higher level may show up as special causes at a lower level. Working from the point of view of the hierarchy has allowed us to apply systems thinking to reduce complexity.

An observer may find that a result seemingly out of control in Production may have as its cause a problem originating in Materials. Looking only at Production as the system establishes the problem outside the system as being a *special cause* outside of Production. You have to deal with "Production's problem" by changing something in the next level Materials system. Looking at the problem and including Materials makes the problem common to the combined, interlinked system of Production and Materials. It's important to put the causes of variation into proper context so that the manager takes appropriate action. When problem causes are found to be coming from outside one level of

	HARD	SOFT
Product Development	Final design	Research, project meetings, layout, testing
Planning	Forecasts, master schedules	Analyzing data, interviewing sales force
Material Flow	Stock on conveyors and shelves	Authorizing purchases, scheduling orders
Production	Nuts, bolts, widgets	Testing, assembling, drilling

Table 9. Examples Within a Hierarchy of Opportunities

hierarchy (from outside the local system), action should be taken at a higher level in the hierarchy. Preventing the cause from recurring will require changing some higher level system.

In our own practice, as we worked with a particular client on their problems, we found the "obvious" problems were in Production. But then we found those problems were actually being driven by the rollover from old to new product: in the transition, either too little or too much of the right products tended to be available to customers. Lack of *materials* was a real problem, and it became clear that better *planning* was needed. We came to realize that planning was being affected by the variability between planned and actual dates when the new products were coming out of *development*. As all this occurred, the understanding ultimately developed that *management* needed to emphasize continuous work on reducing the complexity existing in the entire business system.

Again, the point is that looking at just the Production and Material Flow levels isn't enough. The manager has to deal with all levels of the hierarchy. The number and variety of variables to deal with becomes greater as you move from looking at just the Production level into Materials and beyond. The biggest challenge is that, as you go into higher levels in the hierarchy, selecting the things to measure becomes harder. In addition, it usually takes much longer to collect the data.

Dealing with these challenges simply takes patience and persistence as the "layers of the onion" are eventually peeled back. Also, remember that seeing and dealing with issues that span levels of hierarchy starts with management, not with those on the shop floor.

NOTES

1 John Seely Brown et al., "Situated Cognition and the Culture of Learning," Institute for Research on Learning Report no. IRL 88–0008 (December 1988).

2 Much of this section is in Perry Gluckman's words, describing the core of his work with clients.

3 Management may have read an intriguing magazine article, read a case study in a business review or book on excellence, or used the now popular method of benchmarking to find a "success" to copy.

4 Our associate Joe Reid offers as a counterpoint, "I think no-brainers are easy to find: just talk to the workers who are in contact with the process every day." We leave it to the reader to experiment with these two points of view in their organizations.

5 A common problem with system or process documentation, especially if created fairly rapidly in response to some problem or demand by management, is that it is a mixture of fact and fancy, describing some of what goes on today and some of what the documenters would like to have in future. In short, it describes a system that doesn't exist. For documentation to have value for improving the system it must start by telling the *truth* about the system, as it exists today; adding improvements to the model is a later step.

6 The growing and important activity of software production fits this model of development as well.

What Must Be Done

11.1 Levels of Personal Leadership

Thus far in this book we expect the reader will have seen that the Deming philosophy is about a great deal more than quality of products; it is not limited to business alone. In this section it will become clear as to how Deming's philosophy is about a way of living and of interacting with others, which results in great improvement for businesses, their customers, and profits.

For this reason, Deming's profound changes can't be "installed" as though they were some new program or equipment. Nor is it a matter of executing a procedure of some arbitrary number of steps. When Deming advises top management to "adopt constancy of purpose",[1] he is revealing that pursuit of this philosophy is a long-term effort—something you continue for the rest of your life.

The penalty for not learning Deming's philosophy has been felt by many companies whose "expensive funerals," as Deming calls them, may well be the precursors of a long-term decline in productivity and industrial capacity, possibly coupled with loss of national sovereignty. Nor are we the only ones to see it this way:

> *[Top management] are much like a wealthy family that annually sells acreage. . . . Until the plantation is gone, it's all pleasure and no pain. In the end, however, the family will have traded the life of an owner for the life of a tenant farmer.*
>
> —Warren Buffett, "The Selling of America," *Fortune*, 23 May 1988.

Buffett, by the way, is not a dreamy philosopher or an academic

---

Notes for this chapter begin on page 238.

insulated from the real world: he is a highly successful investment fund manager.

Companies ignorant not only of Deming's profound changes, but also of Packard's equation,[2] often finance their expansion by equity transactions or by going into debt. Such companies—most of Western industry—have increasingly surrendered control of their own affairs to bankers or Wall Street.

As Deming says, "There's a better way." The transformation to his philosophy is something the West *must* do.

Our guide, Perry Gluckman, put it this way:

The message is simple!

1. *Every system is broken.*

2. *Being competitive in the market means improving the system faster than the competition.*

3. *We are all part of the problem, and we are all part of the solution.*

Our message has been directed in particular to Western management, but we offer the challenge that *everyone* can use our recommendations for going forward. Much of it may seem to relate to academicians more than practitioners, but each one of us, regardless of rank or responsibility, can take on tangible tasks. And we will give some examples of those tasks.

If you are a student reading this, find and read other works by and about W. Edwards Deming. If you are writing on continuous quality and productivity improvement, choose topics in the areas of grading, competition, understanding the impact of variation, and so on.

If you are an educator or consultant, reexamine what you teach in light of the Deming philosophy. Study Deming's system of profound knowledge and bring what you offer into accord with it. Guide your students (potential business people, community leaders, and educators themselves) and clients toward application of Deming's philosophy in all aspects of their business and community activities.[3]

For educators, there is even more. Resolve to be a life-long learner. Study, understand, and communicate the process of learning so it can be passed on. Stop grading on the curve—think more of your students than that only a few can get As, and that some have to get Fs.

Parents can make a difference by stopping the encouragement of competition against another person or group of people. Stop offering rewards—they cause the central meaning of a child's effort to become the reward, rather than the effort itself being rewarding. Ask and learn why this would be a better way. Look at the bad long-term effects that can come about, rather than just taking our word for it.

If you are a community leader reading this, apply the principles followed in the work of communities like Madison, Wisconsin; Philadelphia, Pennsylvania; Jackson, Michigan; and Piqua, Ohio. All have formed active councils for improvement, many of them guided by Dr. Myron Tribus.[4] Develop a perspective of local government, business, labor, and education working in concert, all part of the same community system. Understand how it is not at all necessary that an organization be in a "for profit" mode to work actively on improving quality. Invite representatives of the other parts of the community to team together on specific improvement initiatives, guided by systematic review of what to work on.[5]

If you are a member of the community at large, become involved and encourage whatever part of the community you come in contact with to connect with the rest and work on common goals. Donate your particular expertise to the improvement of the community and search for others with the desire to do the same.

These are just a few examples of how Deming's philosophy can be applied in daily life. Another way of seeing your options is to view them as three steps in the transformation of the individual:[6]

1. **Be An Exemplar.** Achieve the degree of understanding of Deming's philosophy that makes it impossible to join in or support the programs of neo-Taylorism. In short, refrain

from wrongdoing. Your life and actions reflect your worldview more powerfully than do your words.

2. **Keep Growing in Knowledge.** Achieve the degree of understanding and exposition[7] required to explain Deming's principles, and point out neo-Tayloristic fallacies to those who come within your own personal orbit. A consistent philosophy requires knowledge.

3. **Widen Your Personal Orbit of Influence.** Achieve the degree of excellence in understanding and exposition that will inspire others to seek you out as a tutor of Deming's philosophy. *As long as you are growing in knowledge, people will seek you out.*

A reconsideration of the words of Leo Tolstoy at the beginning of the book will give you an idea of the potential influence of the "lone" individual undergoing transformation who resolves to be a leader.

The rest of this chapter will focus on the management of business specifically. First will be a brief reminder of how business is too easily influenced by the financial market forces which look for near-term profits at the risk of threatening long-term vitality. This section, and the book, conclude with a summary of where we are today in the world of neo-Taylorism and where we can go through practice of the principles of Deming.

The challenge to Western management is to recognize how our institutions continue to be stuck in "the same old way" of managing—a style that seriously constrains much of the incredible potential to be found within the individuals at all levels who are the ultimate providers of knowledge, initiative, creativity, and true value in our organizations. Will there be a realization of a new paradigm, where the human potential within organizations is fully appreciated and utilized in creating and growing organizations that contribute both to industry and society as a whole?

11.2 A New Direction

Once upon a time it was possible to finance the growth of your business at below the prime rate. Now your friendly banker is more like a loan shark who wants to control your business.

Reducing complexity is the secret to increasing productivity and profitability, and to financing your own growth.

Corporations increasingly are bought and sold today as though they were mere chattel—expedient for the moment, disposed of tomorrow—all in transactions which are perfectly legal. Business prefers the corporate form because of its legal limitation of personal liability for its owners; even professionals such as doctors are increasingly employees of corporations they themselves wholly or partially own.

But the publicly-traded corporation is not only insulated from liability, it is also often estranged from controlling its own affairs. Corporations are creatures of the State: they have no control over who owns them; they are subject to the governance of directors who may have no stake in their future. As expressed as long ago as 1932:

> Economic power, in terms of control over physical assets, is apparently responding to a centripetal force, tending more and more to concentrate in the hands of a few corporate managements. At the same time, beneficial ownership is centrifugal, tending to divide and subdivide, to split into ever smaller units and to pass freely from hand to hand. In other words ownership continually becomes more dispersed; the power formerly joined to it becomes increasingly concentrated. . . .This system bids fair to be as all-embracing as was the feudal system in its time.[8]

Top management, often appointed over the heads of the career employees who know the company and its operations, are responsive to the opinions of stock market analysts and fund managers[9]—a dual constituency whose influence has grown immensely since the above quotation was written. Top management fear these people most of all because to disobey them is to risk the contrived downfall of the company and, along with it, their own investment portfolios. One example of the result of such fear is the practice of borrowing in order to pay the stock dividends expected by the market.

The workers in these companies are especially at risk because they are outside of the power structure, often lack inside information, and work for salaries which allow little accumulation of financial reserves.

Although their mechanism of control differs slightly from that of the stock market, the lenders of money to a company will expect to exert some measure of control, the degree varying with the size of the loan. To borrow without yielding some measure of control is impossible; your creditor moves in as part of your directorship. "Loan-sharking" has traditionally been the lending of money at illegal rates of interest, often with payment enforced by bullies using strong-arm techniques. The bankers who lend corporations large sums of money use much subtler means of control: putting their people on the borrower's board of directors and, as an additional advantage, gaining inside knowledge which they can use profitably in third-party transactions. A lot more is going on than simple borrowing of money; it is a transfer of control.

Companies that learn how to learn and continually reduce complexity will find little need to be under the control of Wall Street or bankers. Adoption of Deming's profound changes by management is the beginning of a company's ability to finance its own growth and thus control its own destiny.

Some time ago Professor Hajime Karatsu of Tokai University told an American reporter:

> It's quite sad, the deterioration of the USA. In five years, you could be No. 1 again. You have all the technology, all the tools. Now if you'd only use them.[10]

We submit that, because of its incalculable leverage on human events, philosophy is by far the most powerful of all tools. Philosophically, Deming has given the West much more than he gave Japan. We ask . . .

When Will the Sleeping Giant Awaken?

11.3 The Restoration of the Individual

We have not yet learned how to live.

—W. Edwards Deming

Part 1.   Neo-Taylorism: Today's Management Philosophy

W. Edwards Deming has said that the central theme of his philosophy is *the restoration of the individual*.[11] Many organizations already claim to have great respect for their individual members. From what kind of treatment or status, then, does Deming want individuals restored?

We believe that neo-Taylorism, with its view of people as bionic machines, is the philosophy of management from which Deming wants people to be restored. It is almost universally practiced today in business, government, and schools, affecting virtually all employed people. Although we have already dealt with many aspects of neo-Taylorism in this book, the few that remain are vital to our discussion:

- Fear
- Internal competition
- Manipulation and control
- The organization over the individual
- "Human Resources"
- Rating and ranking of people

Fear

We have all felt fear. Some fear is healthy—for example, the fear of falling keeps us away from high ledges and precipices, and the fear of being burned keeps our hands away from fire. But the kinds of fear we are illustrating here lead to *un*healthy consequences, for the individual and for the organization:

- Paralysis—reluctance to do things that need to be done
- Anxiety and stress—trying to do the work of a system that does not work, attributing the results to luck in order to avoid confrontation with the real causes

- Cynicism—a defense mechanism born of the desire to be "right," or the numbing attitude that nothing works or ever will
- Passivity—unwillingness to question or to say "no"

These are some of the effects; what are the sources of such fear? Probably the two greatest sources are working under performance-appraisal systems and being made responsible for situations over which one has no control.

We saw examples of both sources in the Red Bead Demonstration, in which people were ranked and rated—and dismissed—on the basis of performance that was entirely determined by the system in which they worked. Management, not they, were responsible for that system. Organizations that use these devices on their people can never gain the rich rewards of cooperation, of people taking joy in their work, of never-ending learning. Deming would say such organizations have no future.

Few are the companies that explicitly describe how they instill fear in their employees. Fear is typically part of the intangible system.[12] Fear also arises in organizations in which there is nothing so obvious as the irascible Foreman Deming—in organizations where management are brilliant and sophisticated. Here the agents of fear may be more subtle, such as in reports to superiors and financial management systems.

The outcomes of fear may be both visible and invisible. Not everyone has the strength of character to remain untouched by fear. Some react by abusing their families or by developing an addiction to alcohol or drugs. Other outcomes are less catastrophic but still damaging:

FOR THE INDIVIDUAL	FOR THE ORGANIZATION
"Fight or flight" behavior	Destruction of teamwork and collaboration
Loss of health	
Mediocrity	Impairment of division of work
Focus on the short term	Mediocrity
Unwillingness to take risks	Decline of communication
Lack of trust	Factionalization
	Lack of trust

Fear's opposite is trust. Trust involves the expectation, belief, and predictability of positive intrinsic motivation or behavior. We are drawn to trust; when we feel trusted we expect fair and just treatment. Trust is also reciprocal: it is received when it is given. Communication—of the type we have described under "New Skills for Managers"—intensifies trust. Deming's philosophy aims to replace fear with trust in all organizations.

Internal Competition

In this book we have many occasions to discuss two different processes of competition, social and free market.

1. **Social Competition** manifests itself in the endeavors of people to court the favor of those in power. This is the type of competition that the first paragraph of Alfie Kohn's book *No Contest—The Case Against Competition* brings to mind:

    > Life for us has become an endless succession of contests. From the moment the alarm clock rings until sleep overtakes us again, from the time we are toddlers until the day we die, we are busy struggling to outdo others. This is our posture at work and at school, on the playing field and back at home. It is a common denominator of American life.[13]

    The conditions and results of social competition are notoriously unpredictable and unfair, with some participants privileged while others are discriminated against. One learns "how to win" in an endless series of fragmented, high-context communications. One of the first rules to be absorbed is that winning is important; another is that the number of winners, and what might be won, are both limited, often arbitrarily, by those in power. Hence cooperation, working together, is artificial and forced; and one can never completely relax one's guard. Your gain is my loss (zero-sum).

    Perhaps the next rule to be learned is that, since the number of winning and runner-up positions is always limited, most of society consists of perennial losers. Even though the rules *guarantee* a large percentage of losers, there is often little compassion for them.

In social competition winners are selected by those in power and the rules they have created, whether the setting be a school, a foundation, or a factory. The participants either cannot, or are unwilling to, change the rules.

2. **Free market competition** is the process by which productive resources come to be deployed in new, more valuable ways, for purposes whose urgency or feasibility had hitherto been overlooked.[14]

The market exists for the purpose of satisfying the desires and needs of consumers. Earlier we cited consumption as the sole object of production; thus, production is one of the essential phenomena of the market.[15] Another is cooperation, both in exchange of goods and services as well as in the division of labor, even among competitors and across national borders.

In the free market there exists a parade of ever-changing opportunities to serve. Searching for, recognizing, and seizing these opportunities is another essential market process which economists call catallactic (exchange) competition. In this system the notion of winners versus losers is inappropriate; consumers, rather than a set of rule-makers, assign everybody a proper place. "Losers" simply find a place in the system that is more modest than what they had planned to attain.[16]

Although Deming is outspoken and explicit against social competition, his approval of free market competition is often implicit and diffuse. Some may read into his call for cooperation among competitors a bias against market competition. Yet market competition is clearly what Deming is urging when he:

- stresses the need for consumer research
- defines quality as something that *increases* the future well-being of consumers
- says that new products and services are generated, among other things, by imagination, innovation risk, trial and error, backed by capital (entrepreneurialism)[17]
- warns us that no one can calculate the future losses of business from a dissatisfied customer
- tells people how to improve quality
- stresses product innovation and improvement
- tells companies they must know their customers' needs better than, and long before, the customers do
- defines production as a system, and tells us that the consumer is its most important part
- says all knowledge (as of consumer needs, of productive capacity) is prediction
- tells companies to "expand the market to meet needs not yet served, then compete"
- warns against merely competing for share of an existing market
- tells producers they are responsible for quality
- points out that consumers did not request the light bulb, pneumatic tire, or other improvements. Entrepreneurs discovered them as opportunities to serve the market.
- warns that one who has vowed to "meet the competition" (for a share of the existing market) has "already lost"
- notes that monopolies may or may not choose to contribute to society, but assures us that monopolies (which he says have the greatest obligation to improve

processes and products) will be overthrown by market
competition if they attempt to fix prices or otherwise fail
to contribute

- tells American management they must learn that in order
  to compete, they must learn to cooperate
- warns us that he who can make a product cheaper can
  take it away from the inventor

Similar to Shewhart's "dynamic process of acquiring knowledge,"
the PDSA cycle, free market competition is an on-going dynamic
process of discovery of the needs of people—another way in which
Deming's philosophy aims at restoration of the individual. Neither
process has any known limit to improvement.

Internal competition within an organization is social competition.
People are forced to compete with one another for money, position,
reward. Departments, divisions, and other units are forced to compete
for prizes, trophies, awards—even for budget allocations. All three
types of systems are involved: formal, informal, and intangible.

When people compete within an organization, it is usually for a
reward that has *artificially* been made scarce. When there are only so
many high performance ratings to go around, *your* getting one affects
negatively *my* chances of getting one. In schools the high performance
ratings are arbitrarily made scarce by grading on the curve. These are
examples of organizational constraint by means of policy. Rarely, if
ever, does anyone ask why a manager would not want people to be so
highly qualified that they *all* get top performance ratings, or why a
teacher would not want *all* students to get top marks. Deming has the
limiting effects of these zero-sum schemes in mind when he exhorts
Americans to adopt whole-system thinking in which "everybody wins."

Does organizational competition bring out the best in us? When
someone else's winning would mean that you must lose, or at least
could not win, you are hardly motivated to help that person do better
or to cooperate. Internal competition is therefore strongly correlated to
the decline of trust among people and organizational units. Similarly,

because internal competition means that some part of the system must lose for another to win, it is also correlated with suboptimization. And just as trust is reciprocated when given, mistrust is also reciprocated; the scene is set for the manipulation of symbols and other behavior which we dealt with at length in section 4.2, "Phases of Learning." This may deteriorate to outright lying—about accomplishments, about costs and schedules, and about other people.

We turn next to internal competition's effects on relationships with superiors and peers. When the next raise, bonus, promotion, or even one's job itself depends upon winning over others, there may be a strong temptation to claim success falsely and to hide problems rather than seek help. To shift attention away from one's own problems one may try to sabotage competitors' efforts by innuendo or false report to management, or by giving misleading or inadequate replies to legitimate inquiries from peers.

Let's not overlook social competition's effects within the person. The prospect of winning—or losing—leads to a narrowing of focus on the prize, the rating, the grade, the "bottom line." The process orientation which marks the desire to improve quality surrenders to a fixation on products and results. Thinking tends to become binary: you are either for me and my project, or you are an opponent. Entrepreneurial risk-taking, which could make one a loser, is increasingly avoided. Creativity and innovation, which might also increase risk, are stifled. People in competition tend toward conformity. They play it safe and perhaps conceal information about their mistakes and failures, information which in an enlightened organization would be a source for learning and improvement.

In the end, if you lose you suffer a loss of self-esteem; if you win you may develop contempt for those who lost—even though the system guarantees there will be losers.

Even where management care nothing for the individual workers, internal competition is still a bad idea because it *undermines interest in the work itself*. If you are competing to win, to prevail, to survive, the job you are supposed to be doing—your work process—must be ranked below the contest itself.

Deming's philosophy offers collaboration as an alternative to zero-sum competition in all organizations. From Latin words meaning "to work with another," *collaboration* is a synonym of *cooperation* and an antonym of *competition*. Within an organization competition, by its nature, cannot achieve whole-system solutions, and it imperils division of labor; these can only be achieved through collaboration. This is why Axiom 1, "Establish leadership and cooperation," precedes the other five axioms. The banishment of internal competition—in view of Deming's goal of restoration of the individual—is the prerequisite to all the rest of his philosophy.

Companies sometimes engage in zero-sum competition. Competing merely for a share of an existing market—so-called "dog eat dog" competition—offers no new advantage to the consumer and does not tend to optimize the system. The better way, as Deming says, is to "expand the market, then compete." Expanding the market means new products and services are created for the consumer.

Organizations also compete in a win–lose context for prizes and awards. Incredibly, one of the most famous is the "quality" award by the U.S. government, the Baldrige Award. Because most competition ignores variation in performance, stressing today's results and ignoring past consistency, one might win the Baldrige today only to fail tomorrow. The Wallace Company of Houston, for example, won a Malcolm Baldrige National Quality Award in 1990, only to find itself sustaining losses a few months later. Giving the lie to Baldrige-winning excellence, Wallace dismissed a quarter of its workforce and filed for Chapter 11 bankruptcy. The U.S. Commerce Department's Curt Reimann averred that Wallace "looked good" on paper and during visits by the Baldrige examiners. Reimann said that his department has "neither the expertise nor the financial resources to perform financial analyses on candidates"—which might reveal that present "excellence" is accounted for by common-cause variation. Wallace's victory (winning a prize) was mistaken for excellence, until its downfall.[18]

Where a free market exists for goods and services—that is, where there are no imposed barriers to entry and no privileged participants—competition is the dynamic process whereby sellers constantly discover newer and better ways to serve buyers by improving the quality and grade of existing products and services, as well as by inventing new ones.[19] Adam Smith observed that this remarkable process operates as though regulated by an invisible hand; today, with the benefit of systems theory, we would say a free market is a self-organizing system using negative feedback mechanisms.

Without the dynamic process of competition, the market system would be static, unlearning, unimproving—exactly the state of affairs Deming told Japanese top management they must change in 1950. Deming's exhortation to "expand the market, then compete" is an invocation of the discovery process called competition. Unfortunately, in English that same word "competition" is used to describe both the beneficial, optimizing, iterative discovery process of the market as well as the damaging, suboptimizing management activity of forcing the elements of an organization to become adversaries of each other.[20] The concepts of social versus market competition offered here may help to clarify those meanings.

Manipulation and Control

In both the business school and the workplace a model belief is that one of management's most critical functions is that of *control:* control of people, control of tasks, control of costs, control over outcomes. In order to gain and maintain control, managers are taught to motivate and manipulate others toward management's ends using a scheme of rewards and punishments, following the blueprint of Skinnerian behaviorism.[21] As we increasingly appreciate the degree to which our lives are externally controlled today, it's easy to see how fear and competition, which we just discussed, fit into this program. We have also encountered this same belief in extrinsic motivation (motivation coming from outside the self) in Taylor's scientific management, where workers were offered money and the prestige of being first-class men if they would meet the work standards set by Taylor's experts. As Deming has said, "There is a better way."

Dovetailing with fear and competition, a system of rewards and punishments is the instrument of control in virtually every organization. Its rubric is "Do this, and you'll get that"—with the corollary, "If you *don't* do this, you'll get something else that you won't like." This can be found in all three organizational systems: formal, informal, and intangible. You have read how the method of dangling rewards and threatening punishments is integral to Taylorism and neo-Taylorism. But how is a manager to know how to motivate any given individual? Psychology comes to the rescue again.

Each tier of an organization has its own hierarchy of needs, according to psychologist Abraham Maslow. These needs can be stimulated by management to provide tailor-made motivation: those at the bottom concentrate chiefly on survival (in Maslow's terms, physiological and safety-related needs) and can be motivated by money. The middle is less motivated by money but can be influenced by position, and "perks" such as larger offices, assigned parking spaces, company cars, and country club memberships (social and esteem-related motivators, per Maslow). At the top, where money, position, and privilege are already enjoyed, senior executives require visible power and prestige, along with the security of golden parachutes, in order to respond to the bidding of their superiors, the investors, and market analysts.

These latter motivators, which Maslow called self-actualization or personal-growth needs, are especially susceptible to manipulation because of their insatiability. No matter how much these needs are satisfied, still more remain to be tapped into as sources for manipulation—which may help to explain why rich and powerful figures, who seem to have everything one could want, appear even more susceptible to scandals than are common people. Another tool is of course fear: along the way, those climbing the ladder of success receive warnings of the punishments for failure, such as exclusion from "club" membership, removal from the fast track, and outright disgrace. "If you can't achieve this goal, we'll put someone in who can . . ."

In this clinical setting, manipulation conveys an element of the grotesque in the world-view of those who advocate the use of goals, quotas, incentives, management by objective, ranking and rating of people, competition, and fear.

Yet direct offers or threats are not always needed. By using slogans and propaganda, management—especially those who practice the manipulation of symbols in order to avoid having to deal with real problems—can sometimes achieve short-term results similar to those possible with the "carrot or stick" approach, without the necessity to reward or punish anyone. Here are some examples:

QUALITY IS UP TO YOU

YOUR JOB IS IN YOUR OWN HANDS

DEFECT-FREE IN '93

DO IT RIGHT THE FIRST TIME

YOU OWN YOUR OWN MORALE

These slogans, as does all propaganda, attempt to create a conditioned response in the hearer or viewer that will result in a given behavior or mind-set when the slogan is repeated.

Rewards and punishments *are* effective—in exacting short-term compliance from their subjects.

If however, your goal is to tap into your employees' intrinsic interest in doing quality work, or to encourage your students to become lifelong, self-directed learners, or to help your child grow into a caring, responsible, decent person, then it makes no sense to ask "What's the alternative to rewards?" because rewards never moved us one millimeter toward those objectives. In fact, rewards actively interfere with our attempts to reach them.[22]

The Organization Over the Individual

We have looked at transactional manipulation and the use of pro-paganda as a means to create patterns of extrinsic motivation toward management-directed goals. Now we turn to some forms that tran-scend the previous techniques, in that their effects are deeper and span longer periods. In some cases they are merely the predominance of the organization over the individual; in others, they amount to psychologi-cal warfare.

An obvious first example is the use of layoffs and other schemes by which management treat the worker as a commodity at their disposal instead of as an inseparable part of the system. In Deming's terms this explicitly places the worker outside the system.

A second and more powerful class of action is the abrogation of promises. A company that has always offered lifetime job security, and has openly relied on this feature to attract new employees as well as to retain experienced ones, announces one day that, inasmuch as these assurances were never legally binding, this practice may be revoked at any time. At the same time a new performance-appraisal system is announced which is clearly designed to make it easy to target any employee for dismissal. After some of the employees leave or are let go, the company may advertise their former positions in the newspaper, hoping to (1) fill them with younger people whose salaries will be lower, or (2) hire back the ex-employees as contractors at a lower salary without benefits, vacation, or opportunities for advancement.[23]

This is a company waging psychological warfare by abrogating its promises. Here are some further examples of abrogated promises:

- Training and retraining can be curtailed, and employees can be forced to pay for their own training.

- Health-insurance schedules can be limited by means of deductibles, lifetime caps, and exclusions.

- Pensions can be weakened by requiring longer time on the job before retirement, and by reducing their dollar value.

- Employees can be forced to contribute where the employer provided the contribution before.

Psychological warfare can also be waged by unilaterally altering relationships: by changing employees from members of staff to suppliers; by putting suppliers, whom Deming says ought to be partners, at arm's length; by changing departments into competing profit centers; and through the creation of hybrid and confusing legal entities, such as making a division into a subsidiary whose management still report to the same management within the now-parent company.

"Human Resources"

Many organizations assert that their employees are their most important assets; others speak similarly about their suppliers. If employees really are assets, then why is this not recognized on the corporate balance sheet? Why is employee education and training charged to an expense account instead of an asset account? Assets are something desirable; why then do some organizations treat their employees as though they were operating expenses—to be reduced as much as possible?[24]

Formerly we had *personnel* departments (note the embedded word "person"); now we have *human resources* departments (note the congruence with the notion of people as bionic machines).[25] In addition, personnel work, including benefits administration, is sometimes contracted out to a third party, leaving employees to deal with disembodied voices at the end of a long-distance phone line instead of fellow members of the organization.

An enabling—and unstated—assumption here is that people cannot be trusted and do not wish to do their best, making it necessary to manage and control them. The organization's loyalty to them is made *conditional*, which ultimately vindicates treating them as liabilities while maintaining the rhetoric of value and respect.

Rating and Ranking of People

Probably no practice of current management is more dogmatically defended *by both its practitioners and victims alike* than that of rating of employee performance—the so-called merit system. The merit system rests on three theories:

1. The theory that people work chiefly for external reward (pay)
2. The theory that it is possible to determine an employee's contribution to the organization
3. The theory that past performance is a valid basis for predicting future performance

Some chapters ago we visited Deming's Red Bead Demonstration. We "watched" as management followed the merit system in motivating and rewarding their employees, and saw how they used employee ratings to halve the size of the organization ("keep the place running with only the best workers") when costs kept overrunning revenue. Yet the result was not success, but failure. What happened?

That people work only, or primarily, for external rewards is a seductively convenient theory, for it leads management directly to the schemes of Skinner and Maslow. Deming dismisses this theory as false and dehumanizing, arguing that contributions and innovations most often arise from people who work in those few systems which allow them to feel they are working for themselves, neither chasing rewards nor in competition with anyone else. People need compensation, of course; but it is perhaps a measure of how impoverished our systems really are that we have little else to dwell on concerning our jobs beyond the rewards they offer. In these systems we are reduced to seeking mere external rewards; but our contributions will be measured and derivative rather than wholehearted and substantive.

Equally deceptive is the theory that contribution to the organization can be determined or evaluated. As we saw in the chapter on Taylorism, the recognition of variation by twentieth century science not only disparages attempts at exact knowledge, but it frustrates with probabilism our desires to predict future events from observed performance. Consider this mathematical representation of an employee's performance:

$$P_x = f(x) + f(y)$$
$$f(y) = f(y_1) + f(y_2) + f(y_3) + \ldots + f(y_n)$$

Here $f(y)$ represents the total influence that the system in which employee x works has on his performance; $f(x)$ represents his own contribution. Clearly the merit system demands that his reward be based on $f(x)$, yet in observing the employee's performance P_x, his manager sees these two factors inextricably confounded—how can she isolate $f(x)$? To make things worse, $f(y)$ is itself a composite of the interaction of many other factors, including the interaction of the employee with the system. To judge an employee's contribution merely from her own observation of his performance (or that of another, such as the customer) the manager must, explicitly or implicitly, assume that the effect of all the many components of $f(y)$—thus is, the total effect of the system on the employee's performance—amounts to zero, clearly an absurdity. How, for that matter, could she even know $f(x)$ if it were the *only* factor?

Even imagining for the moment, however, that management could actually determine the employees' contribution to his own performance, would this be a valid basis for prediction of future performance? Let's return to the Red Bead Demonstration: there management used the employee-contribution values (number of red beads produced) to determine which were the "best" employees. This latter group, however, since the system itself was not changed, performed about as poorly as the original group, leading management to close the plant.

In many organizations one does not have to wait for the annual appraisal to be judged; there are also awards like "Employee of the Month." "Student of the Month" also seems typical, especially in grade schools. One "quality" consulting firm even created a "Least Valuable Employee" award[26] for use on work teams, on the assumption that no one would want to be the one who received it—ignoring the obvious fact that in *any* group someone is always at the bottom, even the group considered the most valuable in the organization. What many people miss, however, is that "most valuable" and "least valuable" awards are wrong for the very same reasons.

In his own, direct way Deming sums up what's wrong with judging people: we're looking in the wrong place!

> *The supposition is prevalent . . . that there would be no problems in production or in service if only our production workers would do their jobs in the way they were taught. Pleasant dreams.*
>
> *The workers are handicapped by the system, and the system belongs to management.*

Management's focus is the future. The present is rapidly becoming the past, and the past is already beyond our control. Management have no valid focus other than the future. They must therefore be able to predict future outcomes with some rational degree of belief; otherwise, their actions will be a random walk, as in rule 4 of Deming's Funnel Experiment. A major lesson of the Red Bead Demonstration is that management can only gain a rational degree of belief in their predictions of the future when they account for the variation inherent in the system and bring the system into statistical control.

Had the management of the Red Bead plant understood that their system was in statistical control, they would have realized there was no way to distinguish one worker's contribution from another's. A constant-cause system was in operation, affecting randomly all workers within it. The effects of the system confounded their attempts to determine individual workers' contributions by tracking their defects.

The merit system is not really about merit at all. The merit of an individual's contribution is impossible to determine, because it is masked by the effects of the system. Even if merit could be determined on some instantaneous basis, the presence of variation renders it useless for predicting future performance, which is management's true concern. Instead of being about rewarding merit, the real function of the merit system is to propagate a myth in order to legitimize the use of fear, competition, and manipulation, and to allow the theories of judgment, reward, and punishment to constrain the future of the organization.

Part 2. Deming's Way: "A New Kind of World"

Throughout this book we've referred to Deming's teachings as a philosophy. Every revolution, whether it be the American Revolution, the French Revolution, the Communist Revolution, good or evil, is preceded by the introduction of a philosophy, or world-view, which gives each revolution its *raison d'être*. These philosophies are usually implanted into society, often decades before, by obscure persons who typically are overlooked—and often even aided—by the very infrastructure whose transformation the ensuing revolution will assure. We saw earlier, for example, that neo-Taylorism is the product of world-views that can be traced back at least two centuries. Similarly, as Deming would point out himself, his own philosophy reflects a philosophical ancestry—one that takes a different branch from that of neo-Taylorism.

The lesson here is that each change that is occurring began years earlier in the mind of some thinker; thus, we must understand and work at the philosophical level if we are to have a desired future effect. If Deming's ideas are to transform the West, it will be a philosophical transformation and not just a technical one.

Just as Deming taught Japanese management in 1950 that the systems they must improve extended beyond their own companies, he teaches today a system view that includes far more than businesses and similar organizations. He is "desperate," he says, for his own country to be guided by this philosophy in its government, its business, its schools—in all aspects—before the decline is too irreparable for America to retain its own national sovereignty.

We dealt with the kinds of things Deming is *against*—the things that destroy people and organizations—in Part 1 of this section. Let's now look at the world as Deming envisions it—a world that would indeed be the result of profound changes!

Joy in Work

One of the most immediately sensed differences in what Deming calls "a new kind of world" would be that people could take joy in their work. Deming's basis for joy in work is not just that fun is better than drudgery; in fact, "joy" here means something deeper than mere fun. One key to joy in work is the security enjoyed by the workers in a Deming-based organization: Deming notes that the term "security" is derived from two Latin words, *se* meaning "without," and *cura* meaning "care." People who do not need to worry whether management will want them around tomorrow are free to contribute to their company.

Deming notes that when people can understand the system in the way depicted by his flow diagram of production viewed as a system (figure 2), they can see who depends upon them in the organization ("why I'm here"), and upon whom they depend. This knowledge helps them to take joy in their work because they can see, or find out, the reasons for doing or producing a particular thing. Other changes which would promote this joy will unfold in the following sections.

Innovation is most likely to occur in people for whom both work and learning are a joy—who feel, as Deming says, they're working for themselves. Here we again see Adam Smith's "invisible hand" at work; because in systems managed according to Deming's philosophy such people are at the same time, and without having to strive consciously at it, serving the aims of the system.

Cooperation

Many aspects of the Deming philosophy are learnt early in life; cooperation is one of these. Cooperation is key to improving the system. Without it we have factionalization with its suboptimizing strategies which waste the system's potential, defeat its goals, and devastate its members. The time to learn cooperation is early in childhood, as an integral part of family membership. One way for parents to teach cooperation is to create an environment in which neither parents nor children compete with one another—in affection and importance, in games, in allocating family resources, in school, at work. Another is to teach the principles of division of labor and unhampered exchange of goods and services that are essential to cooperation in society. Schools must abolish social competition and institute cooperation. Practices that make good grades arbitrarily scarce, such as grading on the curve, must be abolished, along with rules that make an offense of collaboration among students.

Deming's clear intention is that life be lived as much as possible without zero-sum (win–lose) rules. In his thinking, we cannot know how great the sum of a given transaction could be, so why limit it to zero from the start? He has also wisely avoided the intellectual traps of collectivism and central economic planning as ways of withdrawing from competition.

Collectivist societies are not only notoriously poor providers of even basic needs, but also, because they fail to eradicate differences in condition they heighten the causes of envy and political competition for already scarce resources within the system—which is almost always a closed one, kept in existence by other, non-collectivist systems. Many students of American history are aware of the experiments with communalism in the New England colonies in the seventeenth century. Their protracted and costly failure, which resulted in an affirmation of private property and individual responsibility, greatly influenced Thomas Jefferson and through him the fundamentals of the American republic until the end of the nineteenth century.

Another excellent, and probably less-known case is that of Robert Owen, the utopian whom we saw briefly in the section on neo-Taylorism. Owen, who shared the Enlightenment view that man is totally shaped by his environment, set about providing a closed environment that would give "a new existence to man" by surrounding people with superior circumstances only. One of these circumstances was the abolition of private property. His scientific society would "change [society] from an ignorant, selfish system to an enlightened social system which shall gradually unite all interests into one, and remove all causes for contest between individuals" (Owen's words). This "new empire of peace and goodwill" began at New Harmony, Indiana.

By 1826 the community, which had been kept alive by infusions of Owen's private funds, had ground to an economic standstill, and in desperation its members elected three men as dictators. Five months later, however, "Owen was selling property to individuals, and the greater part of the town was resolved into individual lots; commercial enterprises took over most of the stores and sought customers with the vulgar signs of the capitalist heresy . . . and communalism as a way of life vanished as quickly as it had appeared."[27]

No, collectivism is not "Deming's way." He asks, "Is anyone interested in profits?", doubting that most of today's management evidence sufficient interest in running their companies well enough to show a profit.

Another movement is on the rise today which on its face appears to favor cooperation among industry. Its advocates point to the close collaboration in Japan between business, labor, and government in order to sell the idea of emulating it by means of "national economic planning," "national industrial policy," or a "national competitiveness policy" for America, leading to a corporate state comprising a government/business/labor partnership. This is merely collectivism in another form; in fact, government control of the means of production without actual ownership is the chief feature of fascism.[28]

Fascism, Nazism, and the failure of central planning in the Communist people's republics throughout the course of their wretched existence should be powerful lessons for reformers courting the notion of national industrial policy. Pursuit of a "one best way, decided by experts" is just as bad an idea in national affairs as it is in those of a company.

We should realize that power is never created; it is always transferred. When government gains new powers, this means the individual has lost them. Such doctrines as the above, which aim at reducing the freedom of the individual as well as that of the company and the marketplace, clash sharply with Deming's theme of a restoration of the individual.

As we saw in discussing Axiom 1, management must abandon the notion that their job consists of setting goals and controlling outcomes if cooperation is to abide in an organization or in a nation as a whole. Managers who attempt to stage-manage the effects of cooperation without accepting cooperation's deep philosophical basis are merely manipulating symbols for their own gain.

Intrinsic Motivation

Deming's system assumes people work for more than just a reward; that a major component of one's need to work is accounted for by the desire to create something of value. He calls this intrinsic motivation—motivation that is not merely worked up in response to propositions and demands from someone else. Deming has often said that people who enjoy intrinsic motivation feel as though they are "responsible only to themselves" rather than to the management hierarchy.

In his address to the annual Deming User Group Conference in Cincinnati, Ohio, Alfie Kohn offered valuable suggestions to management on how to operationalize the philosophy of intrinsic motivation by ensuring that people's jobs offer them what he called his Three Cs:

- Collaboration—cooperation
- Choice—the freedom to approach a task or problem in a different or unorthodox way, and without fear of the consequences of failure. In Deming's world, failure can be used as a valuable step toward gaining new knowledge.[29]
- Challenge—the opportunity to do meaningful work and solve real and important problems.

What is meaningful work? We might well consider this paraphrase of Walter Shewhart's discussion of quality in the preface to his *Economic Control of Quality of Manufactured Product* (1931):

Broadly speaking, the object of industry is to set up economic ways and means of satisfying human wants and in so doing reduce everything possible to routines requiring a minimum of human effort, through the use of the scientific method, extended to take account of modern statistical concepts.

This might not be the canonical definition of meaningful work and important problems, but it provides an effective practical criterion for assessing almost any activity.

Another example of the recognition of intrinsic motivation is found in the personal mottoes of Soichiro Honda and Takeo Fujisawa, founders of the Honda Motor Company, one of the most highly respected and inventive of the automotive companies:

- Be original
- Do not rely on government
- Work for your own sake[30]

Intrinsic motivation does not mean that pay is insignificant. Deming does play down the importance of pay, understanding that in his system people would tend to be paid what they need. In this need for pay, as in all other needs, people differ. The amount to which pay might be a motivator depends at least partially on what one is making now—another nonlinear relationship calling for knowledge and understanding on the part of management.

System Thinking

Even in this brief discussion of Deming's philosophy it is readily obvious that the parts of it are dependent upon one other for support and effect. One simply can't profitably consider any one apart from the whole—in other words, it is a system. Deming teaches system thinking more by example than by enumeration of its aspects. Although several current writers advocate system thinking, none is so aligned with Deming's philosophy that we can say, "Read so-and-so." However, Peter Senge's *The Fifth Discipline: The Art & Practice of The Learning Organization*[31] offers some advice.

The essence of the discipline of systems thinking lies in a shift of mind:

- *seeing interrelationships rather than linear cause-effect chains, and*
- *seeing processes of change rather than snapshots*

We discussed the notion of nonlinear dynamics in section 2.3; process thinking has, we hope, become obvious as part of Deming's system. We have some closing observations on Deming's system thinking:

- There can be no isolated islands privileged to sidestep improvement within a system
- The system must benefit all of its constituencies, not some at the expense of others
- Everyone, including management, works to improve the system
- Everybody wins
- People desire to do a good job and to contribute
- People enjoy a feeling of being self-employed

Self-Esteem

Self-esteem is best understood in the context of current managerial practice. Consider the Red Bead Demonstration: management applies the system of rewards and punishments typical of most organizations, and the plant fails, with management blaming it on the workers. Without revisiting all the lessons of section 6.7, we note that the Willing Workers did not appear to have much self-esteem. It is this situation Deming is trying to alleviate, rather than putting everyone in love with him or herself.

Part of the resistance to Deming's philosophy is the barrier of ego in people whom the current paradigm of neo-Taylorism has made successful. We surely do not want them to increase their self-love.

For Deming, self-esteem means a midpoint between a concern only for self and the downtrodden archetypes of the willing bead workers. At this midpoint people can develop concerns for wider areas of life, such as the various systems of which they are a part.

In his introduction to *Beyond Culture*, Edward T. Hall states that "many people's sense of worth is directly related to the number of situations in which they are in control." He goes on to say that "many people have problems with their self-image because they are clearly in control of so little." He describes the negative effects (powerlessness, aggression) that suppression of self-esteem can create.

Knowledge Creation

If the methods of W. Edwards Deming could be summed into a single phrase, that phrase would be *never-ending learning*. The uncertain and probabilistic nature of our knowledge does not relegate our enterprises to stagnation, but it does dictate a strategy of constantly gaining new knowledge. For many decades, Deming has stressed the contributions of his mentor Walter Shewhart to business, because Shewhart integrated the scientific method into business operations with what we introduced as the PDSA cycle in section 3.2. With the PDSA cycle any organization can continue to create new knowledge.

New Concerns of Management

In Deming's system, management have many important jobs—almost none of them similar to what they do now:

- Envisioning the long-term future of the system
- Predicting the near-term future by means of statistically based methods and tools such as control charts
- Communicating constantly and effectively with all constituencies in the system
- Giving and receiving trust in every area of the system
- Helping their people (by now we hope it goes without saying that this excludes such travesties as the merit system)
- Being engineers of knowledge, helping their subordinates to transform their knowledge (their "know-how") from its tacit, local, subjective, unquantified, internal state to an external, global, universal, quantified state in which it is usable by an increasingly wider circle within the system

- Constantly improving the system: a job which, as Deming says, "is never finished"
- Widening the scope of the system so as to include more and more of the system of which *it* is a part
- Being concerned with *methods* that more effectively achieve results—not with results alone

It can be seen that in Deming's system the role of the manager is that of servant and guide, not master or hero.

A "New Economic Age"

One of the ways in which Deming's philosophy demonstrates balance and solidity is that, although it focuses on individuals and offers to restore the individual as we described above, it also confers similar benefits upon businesses run (presumably) for profit. These are not two separate concerns: When we respect people as partners, as customers, as neighbors—as fellow members of a system—we tend to treat them accordingly. Shewhart's description of the object of industry can be seen as the result of first having gained a respect for people's wants and needs. All of us have experienced what it's like to deal with producers who do not have this respect!

Many times in this book we've emphasized the human burden of neo-Taylorism and its deadening effects on industry, both of which are visible in everyday life for all to see. Now we will consider quickly the chief *economic* benefits (effects) which would be accessible if Deming's philosophy (the cause) were to be adopted:

- Reduction of economic burden—more for less (the Deming Chain Reaction)
- Expansion of markets
- Survival of organizations that serve consumers (Smith's "invisible hand")

Deming has not been content to prophesy a new economic age if his philosophy is adopted. He has shown us some of the perceptible steps by which it would be achieved, in a model that some call the

Deming Chain Reaction. Shown in figure 3 is an elaboration of Deming's various versions.[32]

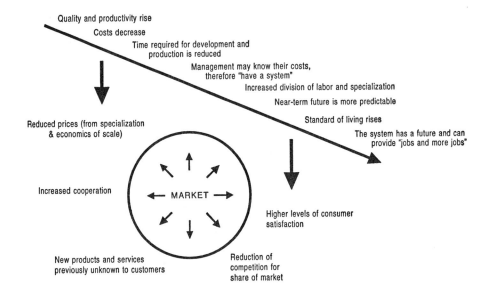

Figure 3. The Deming Chain Reaction.

As we observed at the beginning of this book, unlike some other figures Deming is not a "quality guru," nor should he be associated with symbols such as "total quality management." Quality is an important effect—but only one of the effects—of his philosophy; the additional effects emanate in a vital and growing network of cause and effect, catalyzing the potential for improvement in ways both seen and unseen. Instead of being merely statically divided among competitors, markets are expanded through the dynamic discovery process of competition. And consumers—the only true objects of production—are made happy again and again, by products and services they never even knew to ask for.

The Restoration of the Individual

The promise of Deming's philosophy is nothing less than a much-needed close to the era of Taylorism and neo-Taylorism. The recognition of the individual as part of the system, as a partner—whether a customer, a supplier, a competitor, or merely a neighbor—is long overdue. People are not the "living instruments" and "vital machines" of Robert Owen 160 years ago; nor are they "resources" who can be made the targets of propaganda and exhortation, as they are considered so often today, to be manipulated and then discarded when management tire of having them around.

Deming sees the individual—when not arbitrarily constrained by the system in which he or she works—as having a potential which is, if not unlimited, at least largely unexplored. With the use of Deming's profound changes and profound knowledge, individuals never need stop learning or improving themselves and the systems in which they live.

W. Edwards Deming, not Frederick Taylor, must come to be regarded as "the one relevant social philosopher of this, our industrial civilization."[33]

Whether the West will reverse its already precipitous decline will largely be a function of the dispatch with which its leaders awake and take full responsibility for the consequences of their actions. It will be just as much a function of the sincerity with which they embrace Deming's profound changes as part of their own personal world-views.

We ask again: When will the sleeping giant of the West awaken?

NOTES

1 From his "Fourteen Obligations of Top Management." See Appendix B.

2 Refer again to section 3.3, "Why the Profound Changes Are Important."

3 Several universities are providing Deming-influenced quality improvement programs to students, along with applying the principles to the organization. Two examples are The University of Chicago (contact: William W. Kooser, 312-702-7317) and Fordham University in New York (contact: Dr. Joyce Orsini, 212-636-6219).

4 Refer to discussion of the community improvement council concept in "Creating Community Quality Councils: Applying Quality Management Principles in a Political Environment" by Myron Tribus, Kathy Lusk, and David and Carole Schwinn, a paper based upon presentations at the Hunter Conference, Madison, Wisconsin, April, 1989. Also find the Madison, Wisconsin story in "Quality in the Community: One City's Experience" by George E. P. Box, Laurel W. Joiner, Sue Rohan, and F. Joseph Sensenbrenner (past mayor of Madison) published in *A Practical Approach to Quality: Selected Readings in Quality Improvement*, available from Joiner Associates, Madison, WI, 1989.

5 The papers mentioned in note 4 are good reviews. Even more material is available by contacting the World Center for Community Excellence, 1006 State Street, Erie, PA 16508, (814) 456-9223.

6 Adapted from a lecture at the Foundation for Economic Education by its founder, Leonard E. Read, in 1979.

7 By *exposition* we mean contributing to the body of written literature on the Deming philosophy, as well as giving lectures on various included topics.

8 Adolf A. Berle, Jr., and Gardiner C. Means, *The Modern Corporation and Private Property*, Macmillan, 1932.

9 According to Peter Drucker in *The Post-Capitalist Society* (HarperBusiness, 1992), pension funds in 1991 owned half the capital of the United States' largest businesses, and held almost as much of these companies' fixed debts. These funds do not simply buy and sell stock; they actively wield their power in whatever ways they think will increase dividends and share prices. The largest of the pension funds, the California Public Employees' Retirement System (CalPERS), with investment assets valued greater than $60,000,000,000, is quite open about its attempts to influence the management of the companies whose stock it holds. Even industrial giants like General Motors are quick to listen when CalPERS speaks.

10 Reported in John Hillkirk, "Japan Takes Lead Role on World Stage," *USA Today*.

11 Personal communication, October 1989.

12 A review of Axiom 6 in section 8.2, "Discussion of the Axioms," may be in order.

13 *No Contest—The Case Against Competition: Why We Lose in Our Race To Win*, Houghton Mifflin, 1986, p. 1.

14 We are in debt to Professor Israel Kirzner for this definition and for other valuable advice.

15 "Without someone to purchase our product, we might as well shut down the whole plant." *Out of the Crisis*, pp. 174–175.

16 We recall Adam Smith's "invisible hand," which directs each member of the economic community to produce that which is most urgently required by consumers, the distinctive feature of the system of market exchange being that this "end" need never be consciously aimed at by any participant.

17 *Out of the Crisis*, p. 182.

18 See Jim Smith and Mark Oliver, "The Baldrige Boondoggle," *Machine Design*, 6 August 1992, pp. 25–29. The authors also draw parallels between Wallace's Baldrige adventure and Florida Power and Light's fanatical quest to be the first American company to win a Deming Prize. Although the Deming Prize is non-competitive, FPL's chairman John Hudiberg became obsessed with qualifying for it, and behaved as though the company was in fierce competition to "win" the prize, forcing some employees to work ninety-hour weeks in preparation for the examiners. Even before FPL's achievement was announced, Hudiberg was replaced as chairman by an outsider who dismantled much of what Hudiberg had set up. Competitive behavior may be demanded even where none is called for.

19 Deming often observes that consumers did not request the invention of the telephone, the pneumatic tire, or fuel injection: innovators discovered their needs and provided the products—all part of the on-going process of market competition which makes the producer responsible for quality.

20 Edward de Bono, author of many books on creative thinking, has contributed the term "sur/petition" (see his book of the same title, Harper Business, 1992), which appears to capture some of the discovery process of entrepreneurialism, especially that of improving the grade of product offered. We have not had the opportunity to evaluate the usefulness of this concept in the context of Deming's philosophy.

21 Named after B. F. Skinner, a psychologist who experimented with offering

and denying pigeons food in order to get them to do tricks, and then extrapolated his technique to human beings.

22 From Alfie Kohn, *Punished by Rewards—The Trouble with Gold Stars, Incentive Plans, A's, Praise, and Other Bribes*, Houghton Mifflin, 1993; p. 180.

23 "The Disposable Worker," an article in the *San Jose Mercury News*, 19 July 1993, tells of the "coming of a new social order" pioneered in California's Silicon Valley, featuring the growing phenomenon of disposable contract workers who can be let go at a moment's notice. It cites how temporary employment in 1990 grew 20 times faster than overall employment. Many of the contract workers were, not long before, full-time employees at the hiring firms. Referring to the present state of national health care and the U.S. economy, the president of a local temporary employment agency states, "Let's face it, everyone is going to outsource . . . .What businessman in his right mind would want to hire people?"

24 Even considering people as assets falls short of Deming's ideal, as assets may be disposed of at will; our colleague Dr. Fred Khorasani suggests that the proper relation between employer and employee is that of *partnership*.

25 Some organizations may have made this change with a motive of improvement.

26 Tracy E. Benson, "Quality: If At First You Don't Succeed . . .", *Industry Week* (5 July 1993), p. 58. Our thanks to Dr. Al Viswanathan for bringing this egregious example to our attention.

27 See Chapter 4 of *Can Capitalism Survive?*, by Benjamin A. Rogge, LibertyPress, 1979.

28 See *The Corporative State*, by Dr. Alberto Pennachio; Italian Historical Society, 1927. Nazism differed only slightly, with government actually owning some of the industries it controlled; its version of national economic planning was called the *Zwangswirtschaft*, or compulsory economy.

29 Soichiro Honda, founder of one of the most respected automobile companies in the world, remarked upon his retirement that he felt his career had consisted of "nothing but mistakes, a series of failures, a series of regrets. But I also am proud of an accomplishment." He said, "Although I made one mistake after another, my mistakes or failures were never due to the same reason. I never made the same mistake twice, and I always tried my hardest and succeeded in improving my efforts." Honda, without formal education in science or engineering, understood in a very practical way the application of the scientific method.

30 From Setsuo Mito, *Honda Manejimento Shisutemu (The Honda Book of Management: A Leadership Philosophy for High Industrial Success)*, Athlone Press,

1990. We have insufficient knowledge of how this philosophy is actually applied at Honda, and this citation is, of course, not an attempt to set the company up as an example to follow in implementing the Deming philosophy.

31 Doubleday Currency, 1990, p. 73.

32 See *Out of the Crisis*, p. 3.

33 See the first paragraph of section 2.3, "Taylorism and Neo-Taylorism."

Selected Readings

Order is alphabetical by author within each major section.

Theory of Management

ISO 9000: A Step Backward for the Philosophies of Dr. W. Edwards Deming? Master's thesis by David A. Abell. San José State University, December 1992. Available from UMI Dissertation Services, 300 North Zeeb Road, Ann Arbor, MI 48106-1346.

The Modern Corporation and Private Property, by Adolf A. Berle, Jr., and Gardiner C. Means. Macmillan, 1932. Discusses the transition from stockholders as owners of companies, to stockholders as owners of commodities (their shares of stock), and the resulting divergence of interest between ownership and control (management).

Out of the Crisis, by W. Edwards Deming. M.I.T. Press, 1986. Dr. Deming is one of his own best expositors. This book is not outdated by his more recent book, listed below, and should be read by everyone.

The New Economics for Business, Government, Education, by W. Edwards Deming. M.I.T. Press, 1993. Probably the finest expression of his philosophy to date, bearing the fruits of its continual development throughout the 1980s. The most important single reference work in the field of management today. Available from, among other sources, Quality Enhancement Seminars in Los Angeles, (800) 574-5544, or (310) 824-9623.

A Theory of a System for Educators and Managers (Vol. 21 of the *Deming Library Series*), Films Incorporated, 1993. Dr. Deming and

Dr. Russell Ackoff discuss the properties and approaches of whole-system thinking (synthesis), and contrast it with the typical analytic approach which breaks the system into parts and works on each independently.

My Life and Work, by Henry Ford. Garden City Publishers, 1922. (Republished by Doubleday, 1987.) From the first page Ford demonstrates a seemingly innate knowledge of many of the principles top management have forgotten today. Ford understood, for example, that the way to fund a company's growth is not through loans or stock sales (which puts the company in the hands of financiers and stock market analysts) but by using its productive capacity to provide revenue, and by continually improving efficiency. Thus the company funds its own growth without turning its direction over to outsiders. A timeless work.

"Eliminating Complexity from Work: Improving Productivity by Enhancing Quality," by F. Timothy Fuller. *National Productivity Review* 4 (Autumn 1985): 327–44. A significant portion of work activity is performed just to accommodate the complexity needlessly created by organizational policy. A procedure for identifying such activity is proposed in this article written from Tim's experience in production and quality management positions at Hewlett-Packard.

"How to Construct and Use a Productivity Loss Index," by F. Timothy Fuller. *National Productivity Review* (Spring 1988). This practical paper continues the contribution begun in the above paper, with instructions for demonstrating, in financial terms, how much unnecessary complexity is costing a company.

The Goal: A Process of Ongoing Improvement, by Eliyahu M. Goldratt. North River Press, 1984. The first *quality novel!* Highly realistic and compelling, this book will leave you with some truths and some questions. Excellent illustrations of some of the concepts within the theory of constraints. Goldratt's methods, although not simply a restatement of Deming's, are highly intelligent and quite compatible with them. Available from The Avraham Y. Goldratt

Institute, 442 Orange Street, New Haven, CT 06511, (203) 624-9026.

The Theory of Constraints, by Eliyahu M. Goldratt. North River Press, 1990. The performance of most organizations, Goldratt asserts, is not physically constrained (by manpower or equipment) but is *policy constrained.* If we want to improve productivity, then, we must work on our bottlenecks lest our efforts create capacity which can't be used. Goldratt includes an excellent discussion on the importance and use of the Socratic method in teaching and consulting, and some unusual advice on gaining access to top management in order to teach them. Useful discussion of scientific method. This book is a companion volume to the author's *The Goal.*

Haga's Law, by Dr. William James Haga and Nicholas Accocella. William Morrow, 1980. Why people organize, why organizations don't work, and why, the more we try to fix them, the worse they get. Sound observation and theory, and wide application in all aspects of life.

The Abilene Paradox and Other Meditations on Management, by Jerry B. Harvey. Lexington Books, 1988. The title essay in this group of seven essays on management examines the curious phenomenon of managed agreement. Some of the other essays follow this theme and are well worth studying: "Eichmann in the Organization" deals with the ethics of what would today be called "downsizing." "Group Tyranny and the Gunsmoke Phenomenon" is on maintaining one's principles in the face of "existential risk." "Encouraging Future Managers to Cheat" praises cooperation.

Short-Term America: The Causes and Cures of Our Business Myopia, by Michael T. Jacobs. Harvard Business Review Press, 1991. Excellent chapter on the reasons why owners of corporations have become mere investors trading commodities—and how this has encouraged management to think short term.

The World of W. Edwards Deming (2nd Edition), by Cecilia S. Kilian. SPC Press, 1992. Excellent biographical and background material on the life and career of Dr. Deming, which shows some of the depth of the man. Of special interest are the chapters on his 1950 lectures to Japanese top management, on the requisite qualities of a teacher, and on the contributions of Deming's mentor, the great Walter Shewhart.

No Contest—The Case Against Competition: Why We Lose in Our Race to Win, by Alfie Kohn. Houghton-Mifflin, 1986. "Why does someone have to *lose?*" This is the cry of a youngster, cited by Dr. Deming, who had been having a great time at a Hallowe'en party until someone decided to give prizes for the "best" costumes, and she did not win. *No Contest* is for those who don't yet see why most social competition is harmful—be it competition among siblings, in the classroom, among workers, or among groups. Kohn, a university lecturer in psychology, has united recently with Deming in speaking out against competition. Nor is all the damage psychological; Kohn points out that social competition can preclude the efficient use of economic resources that cooperation allows, such as division of labor.

Punished by Rewards—The Trouble with Gold Stars, Incentive Plans, A's, Praise, and Other Bribes, by Alfie Kohn. Houghton Mifflin, 1993. A reward is anything you are offered on a contingency basis, in order to get you to comply with the offerer's wishes. Reward and punishment, says Kohn, amount to treating people like pets, after the manner of B. F. Skinner—and they are really two sides of the same coin. Far from claiming that rewards are ineffective (they do often get *temporary compliance*), Kohn points out that for any task involving creativity or problem solving, rewards actually decrease performance. Look for extensive citations, background, and references here, as well as a devastating catalogue of reward and punishment's deadly effects in families, in schools, and at work.

"Why Incentive Plans Cannot Work," by Alfie Kohn. *Harvard Business Review* (September–October 1993):54–63. A condensation of some of the arguments from his 1993 book, *Punished by Rewards*.

American Samurai, by William Lareau. Warner Books, 1991. This radical and detailed book, although not pure Deming, is congruent with much of Deming's philosophy and much of what we have said here. According to Lareau, the American Samurai know what American business needs and know what must be done. They do not sacrifice their careers in a futile attempt to lead resistant and irritated sheep out of the darkness. They bide their time, working quietly and secretly to improve the system, until they see that American business is beginning to awake from the stupor of its ignorance. Then, and only then, will the American Samurai come out of hiding in numbers, like Ninja springing from ambush, to work openly to put American business back on top again. Lareau envisions the book as a training course for these Samurai. Well expressed and easy to read, with good summaries wherever needed.

Growth from Performance, by David Packard. Address presented at the Seventh Region Conference of the Institute of Radio Engineers, 24 April 1957. Packard, co-founder of Hewlett-Packard, suggested in 1957 an idea that would be quite novel in today's business environment: *financing a company's growth from its own profits on a pay-as-you-go basis*. Instead of turning the leadership over to financiers and stock manipulators who will control the company for their own ends, top management can, by sound methods and techniques, achieve an adequate rate of profit and capital turnover which can be used to finance a healthy future increase in sales while providing job security and service to the community. Packard even expresses this relationship in an understandable formula. Beyond these unique contributions, Packard also anticipates *kaizen* quality improvement methods and just-in-time manufacturing, which were probably not heard about again in this country for another two decades. This paper is available from the authors.

Driving Fear Out of the Workplace: How To Overcome the Invisible Barriers to Quality, Productivity and Innovation, by Kathleen D. Ryan and Daniel K. Oestreich. Jossey-Bass, 1991. A good antidote to Bardwick's *Danger in the Comfort Zone* (cited in section 7.1, "Excuses Commonly Given for Decline in U.S. Competitiveness"), this book recognizes the harm done by fear in the workplace and offers practical suggestions for driving it out. The tone of this book is clinical; the issues are dealt with pragmatically.

My Years with General Motors, by Alfred P. Sloan. Doubleday, 1964. Sloan, who once said, "General Motors isn't in the business of making cars; it's in the business of making money," also felt that workers have little effect on quality and are generally overpaid. In the book he cites how GM remained profitable through the Depression by laying off tens of thousands of workers. Sloan became famous for instituting centralized control over GM's decentralized operations via powerful staffs and committees, a mode of organization that spread to most other companies. Both his book and his actions establish him as a major contributor to the management philosophy of neo-Taylorism.

Introduction to Quality Engineering: Designing Quality into Products and Processes, by Genichi Taguchi. Asian Productivity Organization, 1986. Dr. Taguchi's exposition of his famous and somewhat controversial ideas, such as the Quality Loss Function, deals with reducing losses due to variability from design through production. The loss function stands in sharp contrast with the so-called "goal-post" notion of quality which assumes that loss begins only when the measured product characteristic is completely outside the tolerance band of the specification. Taguchi was awarded the Deming Prize for his contribution that loss begins with *any* deviation from the nominal value. Some engineering details and formulae.

The Principles of Scientific Management, by Frederick Winslow Taylor. Harper & Bros., 1911. This brief book contains the philosophy that became the chief paradigm of management of the twentieth century, and points to its underlying basis. To understand Deming and his importance, one ought first to understand Taylor, his contributions, and especially the limiting nature of his philosophy.

Ideas Have Consequences, by Richard Weaver. University of Chicago Press, 1948. The practice of management is an effect of the philosophy of management held by managers. In turn, managers choose their world-view, often unwittingly, from those offered them by the thinkers of today and yesterday. Professor Weaver examines several elements of the modern world-view, probes their genesis and development, and evaluates their effects on today's—and tomorrow's—world.

Understanding Variation—The Key To Managing Chaos, by Donald J. Wheeler. SPC Press, 1993. Written specifically for managers, this gem of a book illustrates many ways in which knowledge of Deming's system of profound knowledge would directly and dramatically improve the performance of the theory of variation. One example would be through abandoning the typical binary worldview (in-spec/out-of-spec; OK/not-OK) in favor of recognizing that variation is inherent in all processes and that not all data provide a basis for action. Long an associate of Deming, Dr. Wheeler provides realistic examples to teach each point, and avoids overwhelming the reader with statistical formulae.

Understanding Statistical Process Control (2nd Edition), by Donald J. Wheeler and David S. Chambers. SPC Press, 1992. In his foreword to this book Dr. Deming notes "the advantage of having a process in statistical control. Costs are predictable with a high degree of belief. Limits of variation are predictable." Emphasizing the use of statistics as a tool for learning, prediction, and improvement, the authors take this deep and oft-misrepresented subject and make it as clear and understandable as can be done. This is a practitioner's book that does not shrink from showing us the relevant theory and relating it to the contributions—and often the actual words—of the great Dr. Shewhart.

Principles of Economics

Competition and Entrepreneurship, by Israel M. Kirzner. University of Chicago Press, 1973. Contains excellent contributions to the theory of market competition as an on-going, dynamic process of discovery and fulfillment of consumer needs. Remarkable parallels with Deming's ideas of market expansion, quality improvement, product innovation, and consumer research. Kirzner is a fellow professor with Deming at New York University. His later books include *Discovery and the Capitalist Process* (1985) and *Discovery, Capitalism and Distributive Justice* (1989).

Human Action: A Treatise on Economics (3rd Edition), by Ludwig von Mises. Henry Regnery, 1949. Considered by many to be the most cogent statement of the principles of market economics ever written. In its review, the *Wall Street Journal* commented, "Certainly this book will have great influence if it finds its way where it ought to be, on the bookshelf of every thinking man. Logic may be slow yeast but it works incessantly."

An Inquiry into the Nature and Causes of the Wealth of Nations, by Adam Smith. Originally published in 1776. Widely available today as a standard work in economic theory. Smith clearly states the theoretical case for increasing productivity and quality through division and concurrency of labor, an approach tacitly relied upon by Deming. Also developed is the human basis for division of labor: *cooperation*, a major element in Deming's holistic, win–win philosophy. In the style of its time, the book makes its arguments repetitively, which accounts for some of its forbidding thickness, but most of the above theory is covered in the first six chapters. Essential knowledge!

Knowledge and Learning

The Logic of Modern Physics, by Percy W. Bridgman. Macmillan, 1958. Bridgman, a specialist in high-pressure physics at Harvard, describes the operationalist point of view adopted by Shewhart and Deming. See also C. I. Lewis' *Mind and the World Order*.

Situated Cognition and the Culture of Learning, by John Seely Brown, Allan Collins, and Paul Duguid. Institute for Research on Learning Report no. IRL 88-0008, December 1988.

Toward a Unified View of Working, Learning and Innovating, by

John Seely Brown and Paul Duguid. Institute for Research on Learning (Palo Alto, CA), July 1990.

Applied Chaos Theory—A Paradigm for Complexity, by A. B. Çambel. Academic Press, 1993. Writing for the advanced amateur, the author reveals a charm and modesty that make the difficult parts easier. An excellent overview of a complex (!) subject, with comprehensive treatment of the wisdom and contributions of others.

The Philosophy of Science and Belief in God, by Gordon H. Clarke. The Trinity Foundation (Jefferson, MD 21755) 1964 (Republished in 1987). Critical treatment of the progression of science from Aristotle's "natures" through the mechanism of the Renaissance, and into the probabilism of this century. Emphasis on the philosophy and limits of science. Considerable treatment of Bridgman's work.

A Brief History of Time—From the Big Bang to Black Holes, by Stephen W. Hawking. Bantam, 1988. With an excellent chapter on the reasons for the probabilistic nature of modern physics ("The Uncertainty Principle"), it offers a brief account of Planck's and Heisenberg's contributions. Highly accessible to the layman, reading this one chapter should bring you insight about twentieth century management philosophy and why it had to fail. Brief cameo chapters on Galileo, Newton, and Einstein add personal dimensions to these men, but are unsatisfying in their choice of details.

"The Knowledge-Creating Company," by Ikujiro Nonaka. *Harvard Business Review*, (November-December 1991). One of those rare cases where some of Japan's *real* business secrets are revealed. As with Deming's principles, there's nothing uniquely Japanese about them, and (hint!) *anyone* could use them.

Economic Control of Quality of Manufactured Product, by Walter A. Shewhart. (50th Anniversary Commemorative Reissue) American Society for Quality Control, 1980. Dr. Shewhart, Deming's mentor in the late 1920s at Bell Telephone Labs, was the first to apply statistical control theory to industrial processes. His classic volume of 1931 is delightfully free of late-twentieth-century cant, and provides precepts of quality which are modern even today.

Statistical Method from the Viewpoint of Quality Control, by Walter A Shewhart. Dover Press, 1991. (First published in 1939.) Shewhart's memorable series of lectures on statistical quality control given at the USDA Graduate School in 1938, edited by W. Edwards Deming. A bargain at its original 1939 price, it is now available at about seven dollars from Dover Press. For those willing to study, this medium-sized book offers profound insight into the Deming philosophy. Here we see the beginning of the routine use of scientific method for learning in the PDSA cycle—a significant break with the methods of Frederick Taylor who relied for process improvement almost exclusively on innovative steps initiated off-line by "experts" rather than continuous improvement by workers on the job.

Mind and the World Order: An Outline of a Theory of Knowledge, by Clarence I. Lewis. Dover Publications, 1990. (First published in 1929.) This is a scholarly work in the field of epistemology, a branch of philosophy that deals with knowledge, its limits, and how it is obtained. Dr. Deming refers to it several times in *Out of the Crisis*. Not easy reading, but both Deming and Shewhart found it worth much study. A good companion to Bridgman's *The Logic of Modern Physics*.

Statistical Quality Control—"Thoughtware" for Quality Improvement, by George A. Watson, Ph.D. Paper presented at the William G. Hunter Conference on Quality, Madison (Wisconsin) Area Quality Improvement Network (MAQIN), 10–12 April 1991.

Paradigms and Change

Discovering the Future—The Business of Paradigms (1989). *The Power of Vision* (1990). *The Paradigm Pioneers* (1992). These videotapes produced by Joel Barker provide some excellent material on paradigms and their uses. Barker also has done books and audio tapes.

Science, Order and Creativity, by David Bohm and F. David Peat. Bantam, 1987. A reexamination of Kuhn's theories with new insights, especially on the use of metaphors as aids in the shifting of paradigms. The authors propose that science recover its awe of the universe and concentrate more on ideas than formulae, on the whole rather than fragments. Chaos and order are also discussed. Especially useful for those who are attempting to provoke and direct change.

The Structure of Scientific Revolutions, by Thomas Kuhn. University of Chicago Press, 1970. How paradigms change is the subject of this thoughtful and scholarly book.

Leadership and the New Science—Learning about Organization from an Orderly Universe, by Margaret J. Wheatley. Berrett-Koehler, 1992. Draws parallels between scientific discoveries and their influence on organizations. The author shows the progression from Newtonian to quantum physics and chaos theory, while pointing out how management practices and organizations have changed as a result. She proposes that recent discoveries in science give us an indication of what changes will soon be seen in how we look at work, organizations, and life.

Deming vs. Taylor

Scientific Management in Action: Taylorism at Watertown Arsenal, 1908–1915, by Hugh G. J. Aitken. Princeton University Press, 1985. (First published in 1960.) A detailed and revealing case study of one of the landmark adoptions of scientific management, it provides keen insight into actual implementation steps and the reasoning and planning that preceded them, plus an extended assessment of both the strengths and shortcomings of Taylor's system of management.

The Invisible Powers—The Language of Business, by John J. Clancy. Lexington Books, 1989. A thorough analysis of the use of metaphors by business leaders, and how their choice of language betrays basic attitudes toward the role of management and, indeed, the role of business itself. A key contribution is Clancy's catalogues of entailments— "the relationships and concepts a metaphor brings to mind that determine its power and richness—and can be dangerously inappropriate." Considerable treatment of Taylorism and positivism. An excellent companion for the books on paradigms.

Everyday Heroes of the Quality Movement—From Taylor to Deming: The Journey to Higher Productivity, by Perry Gluckman and Diane Reynolds Roone. (2nd Edition) Dorset House Press, 1993. (First published by SPC Press, 1989.) The association between the late Dr. Gluckman and W. Edwards Deming dates from the mid-1970s. *Everyday Heroes* teaches vital principles by means of a well-known device, the parable. In six parables which deal with the problems, decisions, frustrations, trials, and achievements of people we can really identify with, Gluckman not only brings out many practical applications of Deming's principles, but—and here is perhaps his key contribution—makes the thought processes involved seem simple and obvious. A wise and inspiring book. Also available is the *Everyday Heroes Study Guide,* by Tim Fuller and Marian Hirsch. (Fuller Associates, 200 California Avenue, Suite 214, Palo Alto, CA 94306).

Cultural Aspects

Beyond Culture, by Edward T. Hall. Anchor Books, 1976. Hall develops two important concepts of interest to those trying to make changes in their own cultures: *extensions and extension transfers,* and *context levels.* Extensions are akin to paradigm shifts (see the work of Kuhn, and Bohm and Peat); extension transfers are meaningless or harmful changes made when the true nature of an extension is misunderstood. Context levels refer to the amount of information conveyed (versus implied) in an act of communication, something you'd better be aware of in advocating change. An easily read book with plenty of examples, which will give you new ways in which to think about making changes.

Japan without Blinders—Coming to Terms with Japan's Economic Success, by Phillip Oppenheim. Kodansha International, 1992. One of the few books that deals with industrial Japan as a product of earlier ages, notably the Tokugawa and Meiji eras which shaped Japan's commercial future. Clear treatment of Japan's industry today and of the West's typical attitudes and reactions to it. Not only readable, but fascinating.

"Deming Management Philosophy: Does It Work in the US as Well as in Japan?," by Kosaku Yoshida. *Columbia Journal of World Business* (Fall 1989):10–17. Dr. Yoshida, a frequent lecturer with Dr. Deming, describes the characteristics that the West ought to adopt in large measure if it expects to remain among the world's producer nations.

A much more comprehensive list of books and videotapes which we highly recommend for Deming advocates may be obtained by contacting the author below:

Kenneth T. Delavigne
P. O. Box 13237
Coyote, California 95013

Appendix

A Deming Chronology

Some Important Events in Deming's Life

Our thanks to Dr. Deming for his thorough review of our attempt to capture some of the important events in his long and effective life. When he returned his review copy he had changed all the verbs to the past tense, no doubt because of the intensity of his concentration on the future—of his country and the world. But in order to make the past seem more like the yesterday it really is, we have retained the dramatic present tense.

1900
- Born on 14 October in Sioux City, Iowa.

1903–1906
- Living on the Edwards farm, three hundred acres between Ames and Des Moines, Iowa.

1906
- Deming family moves to Cody, Wyoming. (Wyoming had become the 44th state in 1890.)

1908
- Among first group of settlers in Powell, Wyoming, a barren area made arable by irrigation. The marginal existence shared by everyone engenders in Deming life-long habits of thrift and hatred of waste.

1917–1921
- Attends college at the University of Wyoming in Laramie, supporting himself by doing all kinds of work; earns B.S. degree in engineering.

1921–1922
- Deming teaches in the School of Engineering, University of Wyoming, at the same time studying mathematics.

1922–1924
- Teaches physics at the Colorado School of Mines. In summers he attends the University of Colorado, receiving a master's degree in mathematics and physics in 1924.

1924–1927
- Walter Shewhart, a physicist at Bell Telephone Labs, invents what we know today as the statistical control chart. It allows the variation of the process to be observed by management with radically new insight, making feasible prediction of future performance in a stable process.
- Deming pursues doctoral work in mathematical physics at Yale University, with a part-time teaching position. His Ph.D. dissertation is on the packing of nucleons in the helium atom.

1925–1926
- During two summers Deming works at Western Electric's Hawthorne Works in Chicago, a site for mass production of telephone equipment; during this time the famous Hawthorne Experiments on industrial productivity were run.

1927–1939
- Deming accepts a job at the Fixed Nitrogen Research Laboratory in Washington, D.C. He meets Walter Shewhart, who will have a great effect on his future. Deming visits Shewhart regularly at the Bell Telephone Labs in New York and at his home in Mountain Lakes, New Jersey.

- Beginning in 1930, Deming is a lecturer at the USDA's Graduate School. Over several years he brings as lecturers many of the great statistical thinkers of the day. He writes dozens of papers, at first limited to theoretical physics but later extending to statistical methods.
- In the mid-1930s, Deming spends a year at University College in London on leave from the USDA. He studies under the great statisticians Sir Ronald Fisher, Egon Pearson, and Jerzy Neyman.
- Deming contributes to the growth and understanding of the use of statistics, especially in the area of sampling. His work influences poll-takers, market researchers, and the census.
- In 1931, Walter Shewhart defines an early theoretical basis of quality management and improvement in his book *Economic Control of Quality of Manufactured Product.*

1938–1946

- Deming sponsors Walter Shewhart of Bell Telephone Labs in a four-part series of lectures on statistical control at the U.S. Department of Agriculture Graduate School in Washington (available from Dover Press as *Statistical Method from the Viewpoint of Statistical Control*). Deming also sponsors lectures by J. Neyman, R. A. Fisher, J. Wishart, and William G. Cochran.
- Deming works for the Census Bureau as an advisor in statistical sampling techniques. He begins the use of statistically based survey techniques in the 1940 census, which greatly improve the accuracy and lower the cost of the census.
- Deming also introduces SPC techniques to improve the process of tabulating and summarizing the results (the first use of statistical methods of quality improvement in a white-collar environment?).

- Because of the national interest in census data, this work, combined with what he had done at the USDA Graduate School, greatly increases Deming's esteem in his field.

1942–1945

- The War Department establishes the Office of Statistical Control, recruiting a group of young men known as the Whiz Kids as expediters in moving incredible amounts of war matériel around the world. This group, of which Deming is not a member, sets a precedent for process tampering which will be carried into industry after the war.

- Deming becomes an advisor on statistical education to the Statistical Research Group at Columbia University. He designs curricula based on Shewhart's principles to be taught to people who may have no background in math or statistics, one course for executives and one for workers. Dr. Holbrook Working and Dr. Eugene Grant of Stanford, and Harold Dodge of Bell Telephone Laboratories join this effort. About ten thousand people around the country are educated in these courses. In addition, the Engineering Defense Training Program offers similar courses at the university level.

- Deming lectures on statistics in India.

1945 and immediate post-war years

- The growing movement toward quality created during the war loses momentum as U.S. industry basks for many years in an unprecedented seller's market. Business schools become even more entrenched in teaching a dominant role for finance in management, MBO, etc. Alfred Sloan's and Peter Drucker's ideas are coupled with the least attractive elements of Frederick Taylor's system to form the operative school of American management for the post-war age, neo-Taylorism.

1946

- Deming leaves government service to be a private consultant in statistical studies. He also joins the faculty of New York

University Graduate School of Business where he has taught statistics and management theory until the present day.

1947–1949

- Known for his work in the census, Deming is called to Japan in 1947 by the U.S. Occupation authorities to help Japanese statisticians assess the problems of nutrition and housing in their devastated country and to prepare for a census to be taken in 1951. Unlike other American experts brought to Japan, Deming shows respect for the Japanese people and their culture; he treats them as colleagues instead of vanquished enemies.

- Deming returns to Japan during this period for further advice and collaboration.

1950

- In view of the sad state of Japanese business, Dr. Eizaburo Nishibori, a member of Japanese Union of Science and Engineering, and Professor Sigeiti Moriguti of Tokyo University invite Deming to lecture on statistical methods for business, a session which would be sponsored in July 1950, by the *Keidanren*, the most prestigious society of Japanese executives. Deming hesitates, keenly aware of how rapidly such methods were dropped by American industry. He warns his host, Dr. Ichiro Ishikawa, president of the recently formed JUSE, that unless top management change their school of management both he and Ishikawa would be wasting their time talking to them about quality.

- Ishikawa uses his influence as the former professor of most of the country's top industrialists to assemble forty-five of them for a historic series of conferences in which Deming teaches them the profound changes they must make in their philosophy of management. The executives have little faith in their ability to overcome their past, despite Deming's predictions of success if they follow his teaching, but they begin to do what he said in order to avoid losing face among themselves and with their old *sensei* Dr. Ishikawa. Kenichi Koyanagi, managing director of JUSE, at first skeptical because of the earlier emphasis on inspection as taught by American QC teachers brought in by the Occupation authorities, finds Deming's ideas easy to accept as a basis for hope in a resurgence of Japanese industry. Within ninety days some plants are demonstrating remarkable gains in quality and productivity.

- Deming returns a few months later for further conferences with the same group, and to teach four hundred engineers in eight-day courses.

1951

- Deming returns to Japan three times for further teaching. He also begins to teach sampling methods in consumer research for door-to-door surveys of people's needs.
- JUSE establishes the Deming Prize. It is funded at the start with a donation by Deming of the royalties from his translated books and articles.
- In the period 1950–1970 JUSE teaches Deming's methods to 14,700 engineers and thousands of foremen. In 1986 he will note that these courses are still booked to capacity, as are the courses in consumer research.

1955

- Deming receives the ASQC's Shewhart medal.

1960

- Emperor Hirohito decorates Deming with the Second Order of the Sacred Treasure, an award recognizing the philosophy and methods he provided to Japan which allowed the nation to rise from destruction to the ranks of world economic power. Three decades later, when asked how he felt on receiving this honor, he says he felt "unworthy." "I was lucky," he says, showing he considers himself no exception to his own teachings about the effects of the system upon individual performance.

1980

- On the evening of 24 July, Deming gains belated national attention in his own country for the significant contribution of his philosophy to Japan's amazing business success in an interview in the NBC television "white paper," *If Japan Can, Why Can't We?*
- Begins teaching four-day seminars in the United States.

1982

- Publishes *Quality, Productivity and Competitive Position* (M.I.T. Press).

1986

- Publishes *Out of the Crisis* (M.I.T. Press).
- Elected to the National Academy of Science and Engineering's Hall of Fame

1987

- Defines his system of profound knowledge.
- Begins the videotape series *The Deming Library.*

1988

- Awarded the Medal of Technology by the President of the United States.

1992

- Deming is nominated for the Nobel Prize.

1993

- Publishes *The New Economics for Business, Government, Education* (M.I.T. Press).
- Dies on 20 December in Washington, D.C.

SOURCES FOR THE CHRONOLOGY

Dr. Deming—The American Who Taught the Japanese About Quality, by Rafael Aguayo. Lyle Stuart, 1990.

Out of the Crisis, by W. Edwards Deming. M.I.T. Press, 1986.

The Man Who Discovered Quality: How W. Edwards Deming Brought the Quality Revolution to America—The Stories of Ford, Xerox and GM, by Andrea Gabor. Random House, 1990.

Dr. Eugene Grant. Palo Alto, California.

The Reckoning, by David Halberstam. Morrow, 1986.

Hinshitsu Kanri Kokoroecho (Quality Control Notebook), by Eizaburo Nishibori. Japan Standards Association, 1981.

Appendix

Deming's 14 Obligations of Top Management

W. Edwards Deming created his famous "Fourteen Points" around 1980, and has changed them and added to them several times since then to reflect new knowledge and insight. We can expect this process of refinement to continue; no listing of them will ever be final.

The Fourteen Points are addressed specifically to Western management, the group most in need of Deming's help. "The 14 points are obviously the responsibilities of top management. No one else can carry them out."[1] His later promulgation of his system of profound knowledge does not make the Fourteen Points obsolete; they are still in effect.

Many have mistaken the Fourteen Points as a program or checklist for implementing the Deming philosophy—they are not. The benefits of his philosophy brought out in section 11.3, "The Restoration of the Individual," will not accrue to management who try to transform their organizations "by the numbers."

In 1992, Deming stated, "The fourteen points all have one aim: to make it possible for people to work with joy." This will become apparent as we go through them.

1. **Create constancy of purpose for the improvement of product and service.** Without constancy of purpose in the organization, uncertainty and variability will rise, and parts of the system—employees, suppliers, customers—will no longer be able to work with joy or function productively. Over a

long term, the answers to the questions "What are we
doing?" and "Why are we doing it?" must remain the same.

2. **Learn the new philosophy; teach it to employees, customers, and suppliers; put it into practice.** Although this
   entire book has been about "the new philosophy," we will
   summarize it briefly here as Deming often does: cooperation
   throughout the system, resulting in everybody winning and
   no one losing.

3. **Cease dependence on mass inspection and testing:
   much better to improve the process in the first place so
   you don't produce so many defective items, or none at
   all.** The benefits to the system from this reduction of waste
   of time, money, and effort will be enormous.

4. **End the practice of awarding business on the basis of
   price tag alone; instead minimize total cost in the long
   run.** Understand the advantages—and serious obligations—
   of a long-term relationship with a single supplier, and help
   all suppliers to increase quality and reduce costs. Suppliers
   and customer form a system; optimizing behavior (cooperation) benefits everyone.

5. **Improve constantly every process, whether planning,
   production, or service.** No process is ever "good enough."
   The PDSA cycle, introduced by Walter Shewhart, offers the
   use of the scientific method to gain the knowledge we need
   to improve processes with a high degree of belief that the
   improved level of performance will continue into the future.

6. **Introduce training for skills, taking into account the differences among people in the way they learn.** People are
   the only source of creating improvement and change. If they
   do not grow, continuous improvement will be impaired.

7. **Adopt and institute principles for the management of**

people, for recognition of different abilities, capabilities, and aspirations. The aim here, to quote from *Out of the Crisis*, is not to find and record the failures of men, but to find and remove the causes of failure; to help people do a better job with less effort.

8. **Drive out fear and build trust.** To lead a company focused on improvement requires a different set of values and relationships from one based on control: cooperation is essential throughout the company at every level, based upon respect and trust. Achieving this is purely a function of management.

9. **Break down barriers between staff areas—in other words build a system in which everybody wins.** This means abolishment of internal competition to allow everyone in the system to contribute. When the system wins, everybody in it wins.

10. **Eliminate slogans, exhortations, and targets—asking for zero defects and new levels of productivity.** "Nonsense! You don't need a method. Why weren't you doing it last year? Only one possible answer: you were goofing off."

11. **Eliminate numerical goals and quotas for everybody. They accomplish nothing.** Instead concentrate on what methods are used to perform work, and improve constantly those methods.

12. **Remove barriers that rob people of joy in their work.** This will mean abolishment of the annual rating of performance (the merit system), as well as the grading of students, both of which rank people and create competition within the system. Past performance, even if it could be accurately measured, has little predictive value—and prediction is the job of management. No system can sustain internal competition and remain effective.

13. **Institute a vigorous program of education and self-improvement.** Not to be confused with training (point 6), education gives people the breadth and variety of knowledge to be able to generate and extend ideas for improvement of the system—ideas whose genesis is seemingly unrelated to their immediate jobs.[2]

14. **Accomplish the transformation and continue to study the new philosophy; develop a critical mass in your organization that will bring about the transformation.** The transformation cannot be accomplished by a few people telling others what to do. It requires the cooperation of everyone on a win–win basis.

---

1 W. Edwards Deming, *Quality, Productivity and Competitive Position*, M.I.T. Press, 1982, p. 16.

2 We are indebted to Mr. Earl Werner of General Motors CPC Group for this insight.

Index